Holt
Mathematics

Chapter 3 Resource Book

HOLT, RINEHART AND WINSTON

A Harcourt Education Company

Orlando • Austin • New York • San Diego • London

ISBN 0-03-078298-8

4 5   170   09 08 07

CONTENTS

Holt Mathematics

Date __________

Dear Family,

In this chapter, your child will learn about operations with decimals and fractions and will use decimals and fractions to solve equations.

Your child will first learn to **estimate with decimals**. To **estimate sums and differences with decimals,** use rounding.

Estimate 86.9 + 58.4.

86.9	9 > 5, so round to 87.	87
+ 58.4	4 < 5, so round to 58.	+ 58
	Estimate	145

To **estimate products and quotients with decimals,** use **compatible numbers.** These are numbers that are close to the numbers in the problem and that are easy to multiply or divide.

Estimate 53.62 ÷ 6.45.

53.62 is close to 54. 6.45 is close to 6, and 6 divides 54 without a remainder. **Estimate: 54 ÷ 6 = 9**

To **add and subtract decimals,** line up the decimal points.

Add 3.245 + 9.1.

3.245	Use zeros as placeholders so that both numbers have
+ 9.100	the same number of digits after their decimal points.
12.345	Add each column, just as you would add integers.

In **multiplying decimals,** the product will have the same number of decimal places as the sum of the decimal places in the factors.

Multiply 6 · 0.1.

   6 ← 0 decimal places
× 0.1 ← 1 decimal place
  0.6 ← 0 + 1 = 1 decimal place

Multiply 1.2 · 1.6.

  1.2 ← 1 decimal place
× 1.6 ← 1 decimal place
   72
+ 120
1.92 ← 1 + 1 = 2 decimal places

To **divide decimals,** locate the decimal point in the answer. Then divide as you would whole numbers.

Divide 48.78 ÷ 6.

$$\begin{array}{r} 8.13 \\ 6\overline{)48.78} \\ -48 \\ \hline 7 \\ 6 \\ \hline 18 \\ 18 \\ \hline 0 \end{array}$$

Place the decimal point in the answer directly above the decimal point in the quotient.

If you estimate by dividing 48 by 6, the answer is 8. So, 8.13 is a reasonable answer.

Holt Mathematics

To **estimate sums and differences with fractions or mixed numbers,** you can round the fractions to the nearest $\frac{1}{2}$, then add or subtract. To **estimate products and quotients** with mixed numbers, round to the nearest whole number and multiply or divide.

Adding or subtracting fractions requires a common denominator.

Add $1\frac{2}{15} + 7\frac{1}{6}$.

$$1\frac{2}{15} + 7\frac{1}{6} = 1\frac{4}{30} + 7\frac{5}{30} \qquad \text{Rewrite the numbers using a common denominator.}$$

$$= 8 + \frac{9}{30} \qquad \text{Add the integers and add the numerators.}$$

$$= 8\frac{9}{30}$$

$$= 8\frac{3}{10} \qquad \text{Add. Then simplify.}$$

To multiply with fractions, follow the steps shown below.

Multiply $\frac{1}{3} \cdot 4\frac{1}{2}$.

$$\frac{1}{3} \cdot 4\frac{1}{2} = \frac{1}{3} \cdot \frac{9}{2} \qquad \text{If one of the numbers is a mixed number or a}$$
$$\text{whole number, write it as fraction. } 4\frac{1}{2} = \frac{9}{2}$$

$$= \frac{1 \cdot \cancel{9}^{\,3}}{\cancel{3}_{\,1} \cdot 2} \qquad \text{Simplify.}$$

$$= \frac{3}{2} \qquad \text{Multiply numerators. Multiply denominators.}$$

$$= 1\frac{1}{2} \qquad \text{Rename the product as a mixed number.}$$

To divide **by a fraction,** find the **reciprocal** of the fraction and then multiply. Two numbers are reciprocals if their product is 1. The reciprocal of $\frac{1}{3}$ is 3.

Divide $\frac{2}{3} \div \frac{1}{5}$.

$$\frac{2}{3} \div \frac{1}{5} = \frac{2}{3} \cdot \frac{5}{1} \qquad \text{Multiply by } \frac{5}{1}, \text{ the reciprocal of } \frac{1}{5}.$$

$$= \frac{10}{3} = 3\frac{1}{3}$$

For additional resources, visit go.hrw.com and enter the keyword MS7 Parent.

Holt Mathematics

<table>
<tr><td>LESSON
3-1</td><td>

Practice A
Estimate with Decimals
</td></tr>
</table>

Round to the nearest whole number.

1. 4.23

2. 1.91

3. 10.75

____

____

____

4. 5.88

5. 12.07

6. 18.70

____

____

____

Estimate by rounding.

7. 8.4 + 15.9

8. 3.45 + 5.34

9. 9.36 + 7.542

____

____

____

10. 9.6 + (−7.2)

11. −99.67 + −49.87

12. −4.5 + 6.2

____

____

____

13. 16.8 − 4.3

14. 34.25 − 18.5

15. 23.3 − 3.66

____

____

____

Estimate using compatible numbers.

16. 7.9 • 0.6

17. 6.35 • 1.83

18. 4.4 • 7.15

____

____

____

19. 5.24 • 10.78

20. 29.1 • 5.15

21. 14.25 • 2.99

____

____

____

22. 16.2 ÷ 3.5

23. 32.5 ÷ 3.2

24. 36.34 ÷ 2.1

____

____

____

25. Cheri stopped for gasoline at a station that was
charging $2.39 per gallon. If Cheri had $12.45 in
cash, about how many gallons of gas could she buy? _____________

Holt Mathematics

LESSON
3-1 **Practice B**
Estimate with Decimals

Estimate by rounding to the nearest integer.

1. 7.45 + 35.84

2. 64.08 − 23.47

3. 6.842 + 14.05

4. 7.156 + 8.34

5. 84.23 + (−78.24)

6. 3.78 − 2.078

7. 46.47 − 98.75

8. 87.24 − 56.38

9. 6.324 + 60.324

10. −28.318 + 18.955

11. 35.082 + 8.37

12. −62.49 − 12.84

Use compatible numbers to estimate.

13. 59.69 ÷ 19.904

14. 86.234 • 9.876

15. 54.87 • 19.47

16. −16.04 • 10.45

17. 31.25 • 6.57

18. 92.67 ÷ 32.89

19. 5.548 • 12.38

20. 88.42 ÷ 7.589

21. 90.05 ÷ 6.21

22. Lisha works 20 hours per week at the bowling alley and makes
$8.55 an hour. She gets a raise of $1.30 an hour. Approximately
how much more will she make each week with her raise?

23. Miguel is able to save $87.34 each month. He wants to buy a
guitar that costs $542.45. For about how many months will
Miguel have to save before he can buy the guitar?

4

Holt Mathematics

<table>
<tr><td>LESSON
3-1</td><td>

Practice C
Estimate with Decimals
</td></tr>
</table>

Estimate.

1. $6.187 + 61.871$

2. $52.846 - (-5.284)$

3. $77.435 \div 21.234$

4. $7.894 + (-7.894)$

5. $84.324 \cdot 9.846$

6. $47.445 + 96.845$

7. $-896.23 - 24.571$

8. $-7.501 \cdot 67.499$

9. $483.88 \div (-15.73)$

10. $804.504 + 905.405$

11. $-46.781 - 5.23$

12. $30.214 \cdot 89.321$

13. $98.346 + 162.54$

14. $84.067 + (-101.26)$

15. $720.089 \div 44.506$

16. $645.645 \cdot (-1.49)$

17. $563.647 \div 27.804$

18. $-61.248 + (-65.234)$

19. $2.605 - 11.019$

20. $40.93 \cdot 15.042$

21. $27.38 - (-6.27)$

22. $125.88 \div 23.91$

23. $-38.5 \div 5.34$

24. $59.64 \cdot (-11.6)$

25. Niles is driving from New York City to Baltimore. He drives
64.8 mi/h for 2.9 hours. Approximately how far is Baltimore
from New York City?

____

26. Charlotte had $204. She bought a skirt for $78 and a scarf for
$59. She wants to buy a pair of boots that costs $75. Does she
have enough money left to buy the boots? About how much
money does she have?

____

Holt Mathematics

LESSON 3-1 Reteach
Estimate with Decimals

You can estimate with decimals by rounding each number to its greatest **place**.

Estimate: 52.38 + 9.006 The greatest place of 52.38 is **tens**. The greatest place of 9.006 is **ones**. 　52.38 → 50　　9.006 → 9 　50 + 9 = 59 So, 52.38 + 9.006 is about 59.	**Estimate: 97.45 − 14.9** The greatest place of 97.45 is **tens**. The greatest place of 14.9 is **tens**. 　97.45 → 100　　14.9 → 10 　100 − 10 = 90 So, 97.45 − 14.9 is about 90.
Estimate: 16.35 • 1.8 The greatest place of 16.35 is **tens**. The greatest place of 1.8 is **ones**. 　16.35 → 20　　1.8 → 2 　20 • 2 = 40 So, 16.35 • 1.8 is about 40.	**Estimate: 29.7 ÷ 3.65** The greatest place of 29.7 is **tens**. The greatest place of 3.65 is **ones**. 　29.7 → 30　　3.65 → 4 　30 ÷ 4 is about 32 ÷ 4 = 8 So, 29.7 ÷ 3.65 is about 8.

Estimate.

1. 7.843 + 54.1

Greatest place of 7.843 _________

7.843 rounds to _________

Greatest place of 54.1 _________

54.1 rounds to _________

Estimate: _______________

2. 28.45 − 14.602

Greatest place of 28.45 _________

28.45 rounds to _________

Greatest place of 14.602 _________

14.602 rounds to _________

Estimate: _______________

3. 6.41 • (−19.725)

Greatest place of 6.41 _________

6.41 rounds to _________

Greatest place of −19.75 _________

−19.725 rounds to _________

Estimate: _______________

4. 44.67 ÷ 8.6

Greatest place of 44.67 _________

44.67 rounds to _________

Greatest place of 8.6 _________

8.6 rounds to _________

_________ is about _________

Estimate: _______________

5. 36.5 + 78.09

____

6. 9.45 + (−2.75)

____

7. 98.56 − 53.381

____

8. 33.52 • 5.29

____

9. −68.3 • 4.344

____

10. 24.65 ÷ 4.92

____

Holt Mathematics

Challenge
Just Estimate

Use estimates to compare: $8.62 - 15.91$ ⬚ $-2.4(3.8)$

$$8.62 - 15.91 \boxed{?} - 2.4(3.8)$$

Think: $9 - 16 = -7$ **Think:** $-2 \cdot 4 = -8$

$-7 > -8$, so $8.62 - 15.91 > -2.4(3.8)$

Estimate to compare. Use > or <.

1. $3.45 + 11.65$ ⬚ $16.8 - 4.1$ **2.** $-0.54(14.22)$ ⬚ $-23.65 + 9.44$

3. $-2.59 - 5.18$ ⬚ $6.12 - 13.3$ **4.** $-9.6(-7.25)$ ⬚ $82.45 - 10.3$

5. $24.7 \div 4.5$ ⬚ $12.8 - 5.3$ **6.** $-20.8 + 12.3$ ⬚ $-20.2 \div 1.8$

7. $125.95 - 25.4$ ⬚ $1.5(50.38)$ **8.** $-14.57 - 7.34$ ⬚ $-39.62 \div 2.3$

9. $4.9(-11.6)$ ⬚ $40.29 + 20.51$ **10.** $47.6 - 17.7$ ⬚ $-7.7(-4.1)$

11. $23.6 \div 5.5$ ⬚ $13.08 - 8.4$ **12.** $9.83 - 23.41$ ⬚ $-6.42 - 5.19$

13. $18.45 \div 1.82$ ⬚ $49.5 \div 5.46$ **14.** $0.35(-3.55)$ ⬚ $4.68 - 5.54$

15. $32.7 \div (-2.6)$ ⬚ $-5.28(1.52)$ **16.** $-5.49(-4.06)$ ⬚ $24.62 - 6.4$

17. $-3.24 - 16.4$ ⬚ $-6.6(2.8)$ **18.** $0.78(56.5)$ ⬚ $-9.1 + 66.65$

19. $-45.3 \div 4.55$ ⬚ $79.9 \div (-8.48)$ **20.** $-6.7(-5.08)$ ⬚ $18.5(1.54)$

21. $65.36 \div (-10.7)$ ⬚ $-9.3 + 1.48$ **22.** $-3.53(-2.45)$ ⬚ $-119.5 \div (-11.7)$

23. $65.5 - 67.09$ ⬚ $34.4 \div 33.55$ **24.** $-40.06(-3.1)$ ⬚ $97.8 + 19.7$

Holt Mathematics

LESSON 3-1 Problem Solving
Estimate with Decimals

Write the correct answer.

1. The Spanish Club makes a profit of $1.85 on every pie sold at a bake sale. The goal is to earn $40.00 selling pies. Will the club have to sell more than or fewer than 20 pies to meet the goal?

2. Murray and 4 friends split the cost of a pizza, with each paying the same amount. Murray has $3.36. The pizza costs $14.20. Does Murray have enough to pay for his share? About how much is his share?

3. Luis has 22 MB of free space on his MP3 player. He wants to download 5 songs. They will take up 5.1, 4.1, 4.3, 8.2, and 3.6 MB of space. Does Luis have enough free space? About how much space is needed?

4. Debi has a gift certificate worth $50 for a local book-and-music store. She decides to buy a book that costs $18.95 and some CDs. Each CD costs $13.99. How many CDs can she buy?

Choose the letter for the best answer.

This table shows the number of dollars foreign tourists spent while visiting different countries in 2003.

Spending by Tourists in 2003

Country	Amount Spent by Foreign Tourists ($ billions)
United States	65.1
Spain	41.7
France	36.3
Italy	31.3
Germany	22.8
United Kingdom	19.5
China	17.4
Austria	13.6
Turkey	13.2
Greece	10.6
Mexico	9.5

5. The amount spent in which two countries was about equal to the amount spent in Italy?
   A Austria and Turkey
   B United Kingdom and China
   C Germany and Greece
   D Austria and China

6. About how much did tourists spend in Spain, France, and Italy all together?
   F $111 billion   H $78 billion
   G $109 billion   J $67 billion

7. If the amount spent by foreign tourists remains the same, about how much will foreign tourists spend in the United States over 5 years?
   F $120 billion   H $350 billion
   G $250 billion   J $450 billion

Holt Mathematics

<table>
<tr><td>LESSON
3-1</td><td></td></tr>
</table>

Reading Strategies
Make Generalizations

Estimating is useful when you don't need an exact answer. Knowing how to round decimals helps you estimate.

Rules for Rounding Decimals
1. Look at the digit in the tenths place.
2. If that digit is five or greater, round the number in the ones place up one.
3. If that digit is less than five, round the number in the ones place down one.

Estimate the sum. Use the rules in the table to answer each question.

$$7.48$$
$$+\ 2.68$$

1. In the number 7.48, will you round 7 up or down? Explain.

____

2. In the number 2.68, will you round 2 up or down? Explain.

____

3. Rewrite the equation and solve using the rounded numbers.

____

Use the rounding rules to estimate the difference.

$$13.78$$
$$-\ 5.23$$

4. What is 13.78 rounded to the nearest whole number?

____

5. What is 5.23 rounded to the nearest whole number?

____

6. Is it easier to solve 13.78 − 5.23 or 14 − 5? Explain.

____

Holt Mathematics

Puzzles, Twisters & Teasers

LESSON 3-1

Round and Round We Go!

Estimate sums and differences by rounding. Estimate products and quotients by using compatible numbers. Then solve the riddle.

I −9.916 + 12.4 _______

D 24.79 · 9.83 _______

L 37.2 + 25.83 _______

N −36.8 + 14.217 _______

G 68.2 + 23.67 _______

U 61.45 ÷ 9.08 _______

O 15 − 6.835 _______

R 37.63 ÷ 7.43 _______

A 5.921 − 13.2 _______

H 98.6 + 43.921 _______

E 62.84 − 35.169 _______

Y − 3.96 · 14.81 _______

What did the hat say to the tie?

___ , ___ ___   ___ ___   ___ ___
2   63 63   92   8   8 −23

___ ___ ___ ___ ___ ,   ___ ___ ___
−7 143 28 −7 250   − 60 8   7

___ ___ ___ ___
143 −7 −23 92

___ ___ ___ ___ ___ ___ .
−7   5   8   7 −23 250

Holt Mathematics

LESSON 3-2

Practice A
Adding and Subtracting Decimals

Add. Estimate to check whether your answer is reasonable.

1. 3.52 + 6.33

	Tens	Ones	.	Tenths	Hundredths
		3	.	5	2
+		6	.	3	3
			.		

2. 9.48 + 12.75

	Tens	Ones	.	Tenths	Hundredths
		9	.	4	8
+	1	2	.	7	5
			.		

3. 7.97 + 3.6

	Tens	Ones	.	Tenths	Hundredths
		7	.	9	7
+		3	.	6	
			.		

4. 13 + (−10.35)

	Tens	Ones	.	Tenths	Hundredths
	1	3	.		
+	−1	0	.	3	5
			.		

5. 0.82 + 1.54

6. 12.28 + 19.019

7. 29.5 + 3.676

_______________ _______________ _______________

Subtract. Estimate to check whether your answer is reasonable.

8. 24.6 − 17.93

	Tens	Ones	.	Tenths	Hundredths
	2	4	.	6	
−	1	7	.	9	3
			.		

9. 40 − 19.4

	Tens	Ones	.	Tenths	Hundredths
	4	0	.		
−	1	9	.	4	
			.		

10. 8.4 − 4.6

11. 9.2 − 3.8

12. 13.4 − 8.2

_______________ _______________ _______________

13. 15.42 − 8.3

14. 9 − 5.5

15. 21.68 − 12

_______________ _______________ _______________

11

Holt Mathematics

LESSON 3-2 Practice B
Adding and Subtracting Decimals

Add. Estimate to check whether each answer is reasonable.

1. 6.14 + 8.91

2. 4.51 + 13.08

3. 12.54 + 21.08

____

____

____

4. 34.22 + (−18.5)

5. −10.10 + (−5.9)

6. 6.87 + (−31.6)

____

____

____

7. 9 + 5.68

8. −15.51 + 8.55

9. 36.36 + 54.54

____

____

____

Subtract.

10. 6.23 − 3.62

11. 8.67 − 6.87

12. 28.94 − 9.48

____

____

____

13. 23.57 − 6.84

14. 16.61 − 7.56

15. 32.08 − 12.37

____

____

____

16. 19 − 6.92

17. 42 − 31.89

18. 23 − 21.45

____

____

____

19. 46.2 − 0.27

20. 22 − 18.63

21. 58.9 − 29.58

____

____

____

22. Anna swims the length of the pool in 38.45 seconds and then swims the length of the pool again in 42.38 seconds. What is her total time for 2 lengths of the pool?

____

23. Po has 2 gerbils named Yip and Yap. Yip weighs 3.62 ounces, and Yap weighs 2.79 ounces. How much heavier is Yip than Yap?

____

Holt Mathematics

Practice C
Adding and Subtracting Decimals

Add or subtract. Estimate to check whether each answer is reasonable.

1. $8.326 + 8.49$

2. $13.845 - 10.87$

3. $42.35 + 4.368$

4. $8.695 - 7.52$

5. $-5.421 + 18.4$

6. $-12.54 - 16.951$

7. $6.742 - 9.458$

8. $8.552 + 14.84$

9. $-75.25 + (-6.382)$

10. $47.68 - 9.654$

11. $-38.59 - (-7.816)$

12. $9.561 + (-16.82)$

13. $18.234 + 6.357$

14. $123.45 - 12.345$

15. $6.456 + (-87.51)$

16. $74.832 - 56.842$

17. $95.216 - 59.162$

18. $82.64 + (-45.846)$

19. $27.205 + 19.17 + 8.357$

20. $12.63 + 72.029 + 5.9$

21. $49.408 + 56.27 - 13.02 - 25.821$

22. $9.30 + 17.43 + 38.57 - 25.719$

23. A whippet can run a 200-yard course at a rate of 35.5 mi/h.
This is 3.85 mi/h slower than the top speed of a greyhound.
How fast can a greyhound run?

24. Heike has 2 rocks. If the total weight of both rocks is
54.283 kilograms and the first rock weighs 29.928 kilograms,
how much does the second rock weigh?

Holt Mathematics

Reteach
Adding and Subtracting Decimals

You can use a place-value chart to add decimals.

Add: 3.48 + 2.7

Step 1: Line up the decimal points. Use 0 as a placeholder for hundredths in 2.7.

Ones	.	Tenths	Hundredths
3	.	4	8
+ 2	.	7	0

Step 2: Add. Place the decimal point in the answer.

Ones	.	Tenths	Hundredths
1			
3	.	4	8
+ 2	.	7	0
6	.	1	8

Step 3: Estimate: 3 + 3 = 6. So, 6.18 is a reasonable answer.

Subtract: 68 − 5.9

Step 1: Line up the decimal points. Use 0 as a placeholder for tenths in 68.

Tens	Ones	.	Tenths
6	8	.	0
−	5	.	9

Step 2: Subtract. Place the decimal point in the answer.

Tens	Ones	.	Tenths
	7		10
6	8̸	.	0̸
−	5	.	9
6	2	.	1

Step 3: Estimate to check: 68 − 6 = 62. So, 62.1 is a reasonable answer.

Add or subtract. Estimate to check your answer.

1. 19.67 + 8.45

Tens	Ones	.	Tenths	Hundredths
1	9	.	6	7
+	8	.	4	5
		.		

2. 9.36 − 7.29

Tens	Ones	.	Tenths	Hundredths
	9	.	3	6
−	7	.	2	9
		.		

3. 25.36 − 9.7

Tens	Ones	.	Tenths	Hundredths
2	5	.	3	6
−	9	.	7	
		.		

4. 12.89 + 37.07

Tens	Ones	.	Tenths	Hundredths
1	2	.	8	9
+ 3	7	.	0	7
		.		

5. 34.71 + 9.19

6. 57.06 − 27.38

7. 48.3 + (−10.73)

Holt Mathematics

<table>
<tr><td>LESSON
3-2</td><td></td></tr>
</table>

Challenge
Answer Match

**Add or subtract. Draw lines to match equivalent answers in
each column. Each answer in the middle column matches an
answer in both outside columns.**

1. $-61.244 + 59.014$

2. $21.65 + 14.2 + 10.03$

3. $41.729 + 14.92$

4. $4.018 + 7.052$

5. $212.8 - 136.2$

6. $42.25 - 13.174$

7. $-104.02 + 149.9$

8. $0.015 - 6.72$

9. $-0.673 + (-5.707)$

10. $-5.23 - 0.804 + 35.11$

11. $-70.6 + 48.05 + 33.62$

12. $22.015 + 23.865$

13. $60.25 - 0.601$

14. $65.109 - 5.46$

15. $7.17 - 9.4$

16. $-1.08 + 3.797 - 9.422$

17. $10.25 - 12.03 - 0.45$

18. $16.02 + 44.48 + 16.1$

19. $15.25 + 4.03 + 17.706$

20. $130.82 - 101.744$

21. $-5.61 - 1.095$

22. $11.691 - 21.1 + 3.029$

23. $45.006 + (-8.02)$

24. $52.73 - 53.71 + 12.05$

25. $-33.42 - (-110.02)$

26. $46.01 + 9.9 - 62.29$

27. $6.007 + 74.2 - 43.221$

Holt Mathematics

Problem Solving
Adding and Subtracting Decimals

Write the correct answer.

1. In the Pacific Ocean, the Philippine Trench is 10.05 kilometers deep. In the Atlantic Ocean, the Brazil Basin is 6.12 kilometers deep. How much deeper is the Philippine Trench than the Brazil Basin?

2. Hawaii's Mauna Kea measures 9.75 kilometers from its base to its peak. The base of Mauna Kea lies 5.55 kilometers below the ocean. What is the height of the part of Mauna Kea that is above sea level?

3. A team of mountain climbers makes camp 1.48 kilometers above sea level. They climb another 2.91 kilometers to the mountain's peak. How tall is the mountain?

4. At dawn, the temperature at the summit of a mountain was −8.5°C. By noon, the temperature had increased 3.6°C. What was the temperature at noon?

Choose the letter for the best answer.

This table gives the heights of the tallest mountains on each of the seven continents.

The Seven Summits

Mountain	Country	Height (km)
Mt. Everest	Nepal-Tibet	8.85
Mt. Aconcagua	Argentina	6.96
Mt. McKinley	United States	6.19
Mt. Kilimanjaro	Tanzania	5.90
Mt. Elbrus	Russia	5.64
Vinson Massif	Antarctica	4.90
Puncak Jaya	New Guinea	4.88

5. In 2000, Joby Ogwyn became the youngest person to climb each of the seven summits. How much higher did he climb on Vinson Massif than on Puncak Jaya?

   A 0.02 km   C 0.18 km
   B 0.12 km   D 9.78 km

6. Mt. Kilimanjaro, Mt. Elbrus, and Mt. Aconcagua were the first three of the seven summits Joby climbed. What was the total height he climbed?

   B 16.2 km   H 18.5 km
   G 17.5 km   J 22 km

7. Mt. Aconcagua is about 1.3 kilometers taller than which mountain?

   A Mt. Everest   C Mt. McKinley
   B Mt. Elbrus   D Vinson Massif

Holt Mathematics

Reading Strategies

LESSON 3-2

Use a Grid

A grid is useful for adding and subtracting decimals.

Add 19.2 + 7.54.

Step 1: Make a grid that has enough squares for each number and one for the decimal point.

Step 2: Place one number in each square of the grid. Carefully line up the decimals.

1	9	.	2	
+	7	.	5	4

Step 3: If there is a square without a number, add a zero as a placeholder.

1	9	.	2	0
+	7	.	5	4

Use the grids to answer each question.

1. How do you place the numbers in a grid?

____

2. Why was a zero added to 19.2?

____

3. Write the problem 40.3 – 6.54 in the grid.

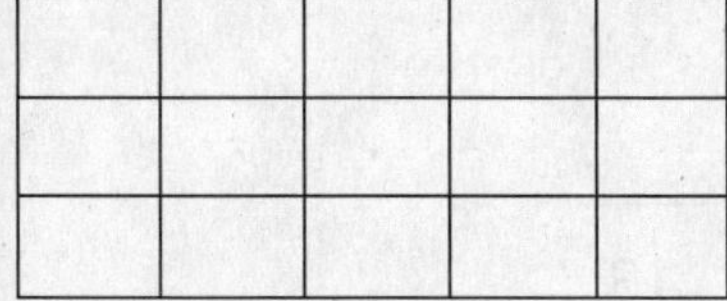

4. Did you need to add a zero as a placeholder? If so, where?

____

5. Subtract 6.54 from 40.30 using a grid. What is the answer?

____

Holt Mathematics

Name ____ Date ___________ Class ____________

Puzzles, Twisters & Teasers

Up Against the Wall!

Add or subtract to find the solution. Round your solution to the nearest integer. Then answer the riddle.

R   5.37 + 16.45   ___________

I   8.89 − 5.91   ___________

E   7 + 5.82   ___________

L   18.31 − 8.66   ___________

N   4.97 − 3.2   ___________

M   7.82 + 31.23   ___________

O   6 + 9.33   ___________

Y   5.98 + 12.99   ___________

C   5.02 + 3.21   ___________

U   5 − 0.53   ___________

H   −7.23 + 6.9   ___________

A   4.16 − 9.04   ___________

T   −32.7 + 62.82   ___________

What did one wall say to the other wall?

___ ___ ___ ,   ___ ___ ___ ___
 3   10 10   39 13 13 30

___ ___ ___   ___ ___   ___ ___ ___
19 15 4   −5 30   30 0 13

___ ___ ___ ___ ___ ___.
 8 15 22 2 13 22

Holt Mathematics

Practice A

LESSON 3-3

Multiplying Decimals

Multiply. Choose the letter for the best answer.

1. 5 • 0.05

  A 25   **C** 0.25

  B 2.5   **D** 0.025

2. 9 • 0.7

  F 63   **H** 0.63

  G 6.3   **J** 0.063

3. 6 • 0.003

  A 18   **C** 0.18

  B 1.8   **D** 0.018

4. 5 • 1.2

  F 60   **H** 0.6

  G 6   **J** 0.06

5. 6 • 1.8

  A 10.8   **C** 0.108

  B 1.08   **D** 0.0108

6. 8 • 0.02

  F 16   **H** 0.16

  G 1.6   **J** 0.016

7. 3 • 8.4

  A 25.2   **C** 0.252

  B 2.52   **D** 0.0252

8. 7 • 0.51

  F 357   **H** 3.57

  G 35.7   **J** 0.357

Multiply. Estimate to check whether each answer is reasonable.

9. 6.8 • 4

10. 8.1 • (−2)

11. 9.5 • 5

12. 3.5 • 7

13. −6.3 • 6

14. 9 • 3.7

15. −6.7 • (−5)

16. 8.8 • (−8)

17. 5.2 • (−4)

18. −3 • 4.1

19. 1.5 • 1.2

20 −2.3 • 1.7

21. Cecile walked 3.7 miles each day for 8 days last month. How many miles total did Cecile walk last month?

Holt Mathematics

Practice B
LESSON 3-3
Multiplying Decimals

Multiply.

1. 6 • 0.3   **2.** 3 • 0.05   **3.** 0.7 • 4

_________   _________   _________

4. 8 • 6.1   **5.** 7.4 • 6   **6.** 1.4 • 9

_________   _________   _________

7. 4.8 • 7   **8.** 3 • 8.2   **9.** 5.5 • 8

_________   _________   _________

10. 1.5 • 6   **11.** 7.9 • 2   **12.** 5 • 6.9

_________   _________   _________

Multiply. Estimate to check whether each answer is reasonable.

13. 6.3 • 7.8   **14.** 9.7 • (−4.7)   **15.** 6.8 • 0.9

_________   _________   _________

16. 2.8 • 8.2   **17.** −7 • 6.42   **18.** 1.9 • 7.22

_________   _________   _________

19. −5.3 • (−8.4)   **20.** 7.16 • 0.03   **21.** 1.56 • (−7.8)

_________   _________   _________

22. 4.6 • 3.1   **23.** 0.62 • 1.45   **24.** −5.74 • 1.9

_________   _________   _________

25. Jordan jogged 4.8 miles each day for 21 days last month.
How many miles did she jog last month?

____

20

Holt Mathematics

Practice C
Multiplying Decimals

Multiply. Estimate to check whether each answer is reasonable.

1. $6.23 \cdot 5.48$

2. $7.82 \cdot 4.18$

3. $9.65 \cdot 4.15$

4. $2.57 \cdot 3.46$

5. $7.84 \cdot 9.23$

6. $4.28 \cdot 6.94$

7. $42.16 \cdot 8.52$

8. $35.84 \cdot 2.57$

9. $10.68 \cdot 71.82$

10. $6.52 \cdot 3.15$

11. $6.78 \cdot (-4.51)$

12. $7.75 \cdot 7.75$

13. $9.49 \cdot 5.25$

14. $-6.84 \cdot 9.44$

15. $10.58 \cdot (-2.54)$

16. $-67.25 \cdot (-5.61)$

17. $94.21 \cdot (-7.81)$

18. $5.27 \cdot (-18.94)$

19. $-1.05 \cdot 3.8 \cdot 7.1$

20. $0.45 \cdot 1.8 \cdot 0.06$

21. $5.7 \cdot 4.52 \cdot (-8.3)$

22. $0.05 \cdot 1.7 \cdot (-12.62)$

23. $4.02 \cdot (-9.9) \cdot 23.7$

24. $-2.2 \cdot 18.45 \cdot (-0.8)$

25. Phone service costs \$24.00 per month plus \$0.09 per
minute for long distance calls. What is your monthly bill
if you make 57 minutes of long distance calls? _______________

26. A soccer uniform costs \$3.98 for the shirt and \$4.49 for
the shorts. What is the total cost of 5 uniforms? _______________

Holt Mathematics

LESSON 3-3 Reteach
Multiplying Decimals

To multiply two decimals:
Step 1: Round each number to the nearest integer.
Step 2: Multiply the integers to estimate the product.
Step 3: Multiply the decimals.
Step 4: Place the decimal point in the product
to make it closest to the estimate.

Multiply: $2.7 \cdot 4.3$

```
   4.3
   2.7
   301
   860
 11.61
```

11.61 is close to 12.

Think:
2.7 rounds to 3.
4.3 rounds to 4.
$3 \cdot 4 = 12$
Place the decimal point in the product to make it closest to 12.

Multiply.

1. $6.7 \cdot 9.1$

6.7 rounds to _______________

9.1 rounds to _______________

The product is close to __________.

Product: _______________

2. $-3.21 \cdot 8.8$

−3.21 rounds to _______________

8.8 rounds to _______________

The product is close to __________.

Product: _______________

3. $4.1 \cdot 0.8$

4.1 rounds to _______________

0.8 rounds to _______________

The product is close to __________.

Product: _______________

4. $12.3 \cdot (-2.7)$

12.3 rounds to _______________

−2.7 rounds to _______________

The product is close to __________.

Product: _______________

Multiply. Estimate to place the decimal point.

5. $2.06 \cdot 7.9$

6. $-4.89 \cdot 0.6$

7. $8.23 \cdot (-4.2)$

_______________   _______________   _______________

Holt Mathematics

LESSON 3-3 Challenge
Decimal Round

Think about numbers that could have resulted in this estimate.

$$0.8 \cdot 3 = 2.4$$

The *least possible factors* that could have been rounded up are 0.75 and 2.5. So, the *least product* is $0.75 \cdot 2.5 = 1.875$.

The *greatest possible factors* that could have been rounded down are 0.84 and 3.4. So, the *greatest product* is $0.84 \cdot 3.4 = 2.856$.

Find the products of the rounded numbers below. Write the least and greatest factors for each product before rounding. Then find the least and greatest products.

	Rounded Product	Least Product	Greatest Product
1.	$18 \cdot 12 =$		
2.	$3.3 \cdot 4 =$		
3.	$0.9 \cdot 0.7 =$		
4.	$1.8 \cdot 2.2 =$		
5.	$6 \cdot 0.1 =$		
6.	$4.5 \cdot 0.6 =$		
7.	$10 \cdot 1.1 =$		
8.	$0.3 \cdot 15 =$		
9.	$8.7 \cdot 5.1 =$		
10.	$1.4 \cdot 1.9 =$		

Holt Mathematics

Problem Solving

LESSON 3-3

Multiplying Decimals

Write the correct answer.

1. A group of 6 adults bought tickets to a play at the community center. The tickets cost $17.50 each. How much did the tickets cost in all?

2. Student tickets for the basketball playoffs cost $9.25 each. How much do 9 student tickets cost?

3. Movie tickets for senior citizens cost 0.8 times as much as regular adult tickets. Adult tickets cost $7.50. How much do senior citizen tickets cost?

4. About 30.5 million Americans attend classical music concerts. The average concertgoer attends 2.9 concerts per year. About how many tickets to classical concerts are sold each year?

Choose the letter for the best answer.

5. The population of Illinois in 2000 was about 1.57 times its population in 1940. If the population of Illinois in 1940 was about 7.9 million, what was its population in 2000, to the nearest tenth of a million?

   A 1.2 million

   B 9.5 million

   C 12.4 million

   D 15.8 million

6. In 2004, the United States had 32.5 million broadband subscribers. This number is expected to increase 1.75 times by 2008. How many broadband subscribers is the United States expected to have in 2008?

   F 24.375 million

   G 34.25 million

   H 56.875 million

   G 568.75 million

7. Nauru and Gibraltar are among the smallest countries in the world. Nauru has about 3.28 times the area that Gibraltar does. Gibraltar is 2.5 square miles. What is the area of Nauru?

   A 0.78 mi^2

   B 0.82 mi^2

   C 5.78 mi^2

   D 8.2 mi^2

8. From 1999 to 2008, the United States Mint is issuing a series of 50 quarters representing each of the 50 states. What is the cost of collecting 10 of each quarter?

   F $100

   G $125

   H $150

   J $250

Holt Mathematics

Reading Strategies

LESSON 3-3 *Compare and Contrast*

Decimals are multiplied in much the same way that you multiply whole numbers.

Multiply Whole Numbers	Multiply Decimals
5	0.5
× 7	× 0.7
35	0.35

Compare multiplying whole numbers to multiplying decimals.

1. What is the same about multiplying whole numbers and decimals?

   ____

2. What is different about multiplying whole numbers and decimals?

   ____

   ____

It is important to place the decimal point correctly in the product.

Steps for Placing the Decimal Point in the Product	Example: 1.37 × 0.8
Step 1: Find the product.	1096
Step 2: Count the number of decimal places in each factor.	1.37 $\longrightarrow$ 2 places 0.8 $\longrightarrow$ 1 place
Step 3: Find the total number of decimal places in both numbers.	3 places
Step 4: Using the number found in Step 3, move that number of places to the left in the product and place the decimal point.	1.096 3 2 1

3. How many decimal places are in 0.63?

   ____

4. How many decimal places are in 4.231?

   ____

5. How many decimal places will be in the product of 0.63 × 4.231?

   ____

Holt Mathematics

Puzzles, Twisters & Teasers

LESSON 3-3

Turning of the Times!

Multiply the decimals to solve. Round your solution to the nearest integer. Then answer the riddle.

S 9 • 0.36 _________

O 8 • 0.6 _________

U 6 • 4.9 _________

T 2.4 • 3.2 _________

I 3.0 • 3.2 _________

R 0.6 • 0.05 _________

N 3 • 7.01 _________

A − 2.6 • 0.4 _________

G 8 • 0.77 _________

H 1.7 • 12 _________

E 2.8 • −1.6 _________

W 7 • 0.3 _________

When is a car not a car?

___ ___ ___ ___   ___ ___
2   20 −4 21   10 8

___ ___ ___ ___ ___
8   29 0   21 3

___ ___ ___ ___   ___
10 21 8   5   −1

___ ___ ___ ___ ___ ___
6   −1 0   −1 6   −4

Holt Mathematics

Practice A
Dividing Decimals by Integers

Match each division with the correct quotient.

1. $4.5 \div 3$	**A.** 0.8	**9.** $2.79 \div 3$	**R.** 2.4
2. $6.4 \div 8$	**B.** 1.21	**10.** $9.45 \div (-7)$	**S.** 0.93
3. $8.1 \div 9$	**C.** 1.15	**11.** $21.6 \div 9$	**T.** 12.3
4. $4.84 \div 4$	**D.** 1.5	**12.** $-2.5 \div 5$	**U.** -1.35
5. $1.68 \div 8$	**E.** 0.9	**13.** $24.6 \div 2$	**V.** 0.59
6. $5.04 \div 7$	**F.** 0.91	**14.** $7.7 \div (-11)$	**W.** -1.31
7. $6.9 \div 6$	**G.** 0.21	**15.** $5.9 \div 10$	**X.** -0.5
8. $4.55 \div 5$	**H.** 0.72	**16.** $-10.48 \div 8$	**Y.** -0.7

Divide. Estimate to check whether each answer is reasonable.

17. $5.4 \div 9$ **18.** $0.66 \div 3$ **19.** $15.3 \div (-6)$

_______________ _______________ _______________

20. $-2.8 \div 14$ **21.** $13.04 \div 8$ **22.** $-5.5 \div 11$

_______________ _______________ _______________

23. Alyssa and Kim bought a new volleyball for $12.95 and
a new net for $16.75. They shared the cost equally.
How much did they each pay? _______________

24. Dominic's scores for his skating performance were
5.5, 5.2, and 5.5. What was his average score? _______________

Holt Mathematics

Practice B

Dividing Decimals by Integers

Divide. Estimate to check whether each answer is reasonable.

1. $20.8 \div 8$

2. $54.4 \div 5$

3. $0.876 \div 6$

4. $65.6 \div 4$

5. $-96.88 \div 7$

6. $50.4 \div 18$

7. $67.42 \div 4$

8. $88.65 \div (-3)$

9. $77.25 \div 5$

10. $-0.18 \div 4$

11. $41.17 \div (-23)$

12. $74.55 \div 25$

13. $0.144 \div 4$

14. $5.36 \div (-8)$

15. $27.6 \div 12$

16. $22.08 \div (-3)$

17. $1.976 \div 13$

18. $25.56 \div (-5)$

19. $0.504 \div 9$

20. $170.1 \div 27$

21. $5.25 \div (-3)$

22. Doris collects wicker baskets. She spent $9.56 on 3 baskets at the flea market. Then she found 4 more baskets at a garage sale. She paid $10.67 for those baskets. What was the average price per basket for all 7 baskets?

23. As of January 2002, 3 top college football coaches had the following winning percents in bowl games: 0.740, 0.692, and 0.683. What was their average winning percent?

Holt Mathematics

LESSON 3-4 **Practice C**
Dividing Decimals by Integers

Divide. Estimate to check whether each answer is reasonable.

1. $86.24 \div 8$

2. $56.64 \div 4$

3. $0.735 \div 7$

4. $75.75 \div 5$

5. $92.7 \div 9$

6. $243.25 \div 7$

7. $83.61 \div 3$

8. $312.85 \div 5$

9. $-0.846 \div 18$

10. $67.656 \div 12$

11. $395.55 \div (-15)$

12. $10.353 \div 21$

13. $581.98 \div (-14)$

14. $-291.89 \div 17$

15. $122.21 \div 11$

Simplify each expression.

16. $2.8 - 14.07 \div 3 + 6.2$

17. $26.65 \div 4.1 + 8.047$

18. $20 \cdot 32.4 \div 6 \times 5$

19. $1.22 \cdot 1.1 \div (-2) - 0.57$

20. $(91.89 + 10.51) \div 16 + 1.045$

21. $67.44 \div 24 + 0.23 - 14.076$

22. An investor pays $478.75 to purchase 25 shares of stock. The total price includes a $35 commission. What is the cost of each share of stock? _________

23. Quinn swam three laps during her swimming class. If her times were 32.02 seconds, 33.48 seconds, and 32.96 seconds, what was her average lap time? _________

Holt Mathematics

Reteach

LESSON 3-4

Dividing Decimals by Integers

You can use models to divide decimals by integers.

Divide 0.35 ÷ 5.

Represent 0.35 by shading 35 out of 100 squares.

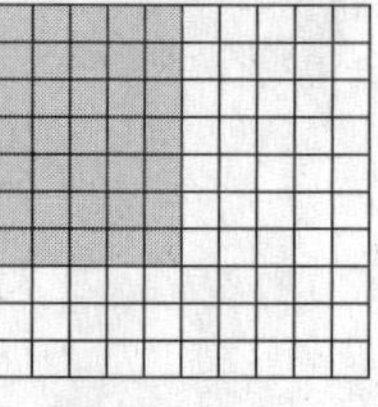

Divide the shaded area into 5 equal parts.

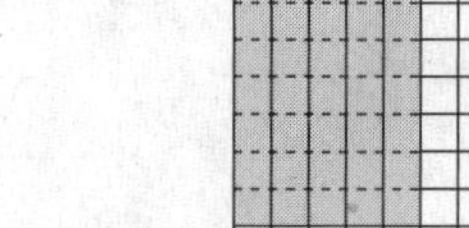

Each part is made up of 7 squares. This represents 0.07.

$$0.35 \div 5 = 0.07 \text{ or } 5\overline{)0.35}$$

Notice that the decimal point in the quotient is directly above the decimal point in the dividend.

Use the model to divide.

1. 0.48 ÷ 4

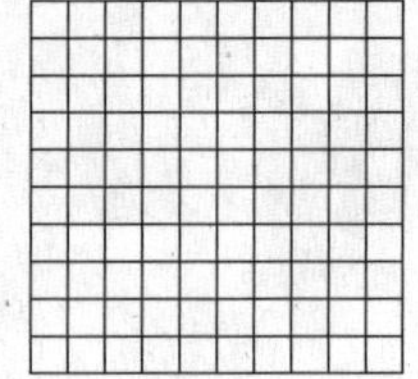

____

2. 1.8 ÷ 3

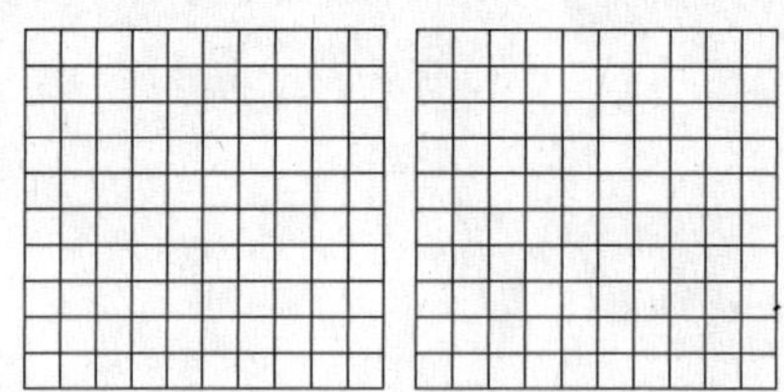

____

3. 0.16 ÷ 4

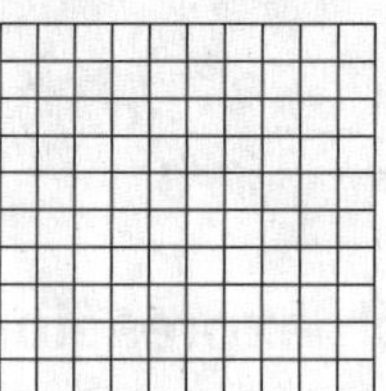

____

Divide. Draw models to help you.

4. 0.56 ÷ 8

____

5. 0.75 ÷ 5

____

6. 7.2 ÷ 9

____

7. 10.5 ÷ 5

____

8. 6.4 ÷ 16

____

9. 7.08 ÷ 6

____

Holt Mathematics

LESSON 3-4 — Challenge
Code to Order

Simplify each expression. Write the answers in order from least to greatest on the top lines of the certificate below. Then write the letter that corresponds to each answer in the same order on the bottom lines to see a secret message.

1. $5.43 + 8(6 - 2.7) - 0.3^2 \div 9 + 12.6$ _______________ [O]

2. $4^2(2 - 0.5) - 6.08 \div 2^3 + 13 \cdot 4.2$ _______________ [O]

3. $(7.4 + 2.1) \div 5 + 1.2 \cdot (9 \div 0.3) - 8.4$ _______________ [E]

4. $126 \div 12 \cdot 2 - 8(0.2 + 5) \div 2$ _______________ [A]

5. $4 \div 0.2^3 \div 5 + (-5.95 + 5.45)^2$ _______________ [B]

6. $72.32 \div 4 - 0.2^2 + 2.1 \div 0.3$ _______________ [W]

7. $(0.2 \cdot 40 \div 2^2 + 10) \div 0.2^2 \cdot 0.5^2$ _______________ [J]

8. $8(32 \div 1.6) \div 0.8 \cdot (0.7^2 + 0.01) \div 2.5$ _______________ [S]

9. $54.18 + 2.34 \div 3.9 - 2.1 \cdot 1.2$ _______________ [M]

10. $8.35 \div 5 + 5(4.86 \div 1.8)^2 \cdot 2$ _______________ [E]

Holt Mathematics

LESSON 3-4

Problem Solving
Dividing Decimals by Integers

Write the correct answer.

1. Teri collects loose change in 3 cans placed near cash registers at the mall. One can holds $37.18. The second can holds $44.25. The third can holds $50.84. If Teri divides the money equally among 3 charities, how much money will each charity get?

2. In 2003, listeners in the United States paid $0.07 billion to download music. The expected increase in sales of downloaded music is $2.19 billion by 2008. What is the average increase in sales expected over each of those 5 years?

3. The longest life expectancies are found in Okinawa (81.2 years), Japan (79.9 years), and Hong Kong (79.1 years). Life expectancy in the United States is 76.8 years. About how much lower is that than the average life expectancy in the top 3 locations, to the nearest tenth?

4. The Ralstons are driving from Washington, D.C., to Los Angeles. The trip is 4,234.185 kilometers. The Ralstons want to make the trip in 5 days, driving the same distance each day. About how far will they drive each day? Round your answer to the nearest whole number.

Choose the letter of the best answer.

5. A top-grossing rock tour sold $121.2 million worth of tickets over the course of 60 shows. What was the average value of the tickets sold at each of the shows?

  A $20.2 million

  B $2.02 million

  C $0.202 million

  D $7.27 million

6. A pizza parlor sells an 8-slice pizza for $9.20. If the pizza parlor charges a total of $1.20 more for a pizza it sells by the slice than for a pizza pie it sells whole, how much does a slice cost?

  F $1.15

  G $1.25

  H $1.30

  J $1.50

7. Bill orders 3 copies of a book from a book club. The cost is $8.00, which includes $0.50 for handling. How much does the book cost?

  A $2.66   **C** $0.25

  B $2.50   **D** $4.00

8. An apple pie has 27.6 grams of fat. It is cut into 6 slices. How many grams of fat are in 2 slices?

  F 4.6 grams   **H** 13.8 grams

  G 9.2 grams   **J** 92.0 grams

Holt Mathematics

LESSON
Reading Strategies
3-4 *Use a Flowchart*

Dividing a decimal by an integer is like dividing by a whole number.

| **Step 1:** Place a decimal point in the quotient above the decimal point in the dividend. | **Step 2:** Divide as you would with whole numbers. | **Step 3:** Estimate to check to see if the answer is reasonable. |

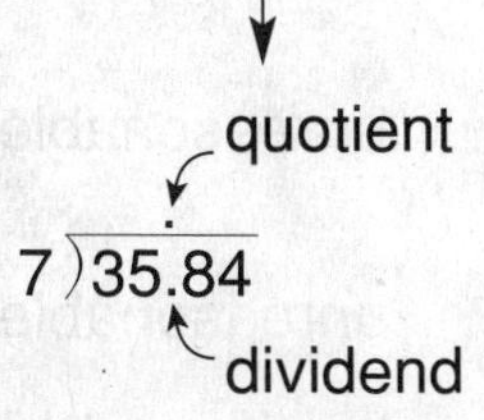

$$7\overline{)35.84}$$

quotient / dividend

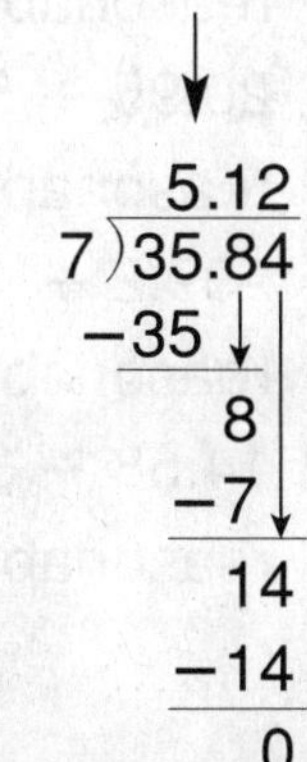

```
      5.12
7)35.84
 −35
    8
   −7
   14
  −14
    0
```

$$35 \div 7 = 5$$

Use the flowchart and the problem above to answer each question.

1. Where should you place the decimal point in the quotient?

____

2. Do you think the estimate in the problem above is reasonable? Why or why not?

____

Use the flowchart and the problem $48.5 \div -5$ to answer each question.

3. Show where the decimal point should be placed in this problem.

$$25\overline{)48.5}$$

4. Divide. What is the answer?

____

5. Is the answer reasonable? Why or why not?

____

Holt Mathematics

LESSON 3-4 Puzzles, Twisters & Teasers
R. U. Serious?

For each equation shown, decide whether the given solution is reasonable or unreasonable and circle your answer. Then use the Secret Decoder to solve the riddle. You'll need to choose one letter not to use.

1. $42.98 \div 7 = 600.14$

   reasonable   unreasonable

2. $-94.72 \div 37 = -2.56$

   reasonable   unreasonable

3. $12.8 \div 4 = 32$

   reasonable   unreasonable

4. $21.47 \div 19 = 1.13$

   reasonable   unreasonable

5. $0.148 \div 4 = .0037$

   reasonable   unreasonable

6. $65.28 \div 32 = 2.04$

   reasonable   unreasonable

7. $0.84 \div 12 = 7.12$

   reasonable   unreasonable

8. $20.95 \div 5 = 4.19$

   reasonable   unreasonable

9. $-39.2 \div 14 = 2.8$

   reasonable   unreasonable

10. $14.58 \div 3 = 4.86$

    reasonable   unreasonable

Secret Decoder

Reasonable	Unreasonable
D J	M B
N A O	P S E

Who is the most famous underwater spy in the world?

__   __   __   __   __
R   R   U   U   U

__   __   __   __
U   R   R   R

Holt Mathematics

<table><tr><td>**LESSON 3-5**</td><td># Practice A

Dividing Decimals and Integers by Decimals</td></tr></table>

Divide. Estimate to check whether your answer is reasonable.

1. $7.5\overline{)15}$ **2.** $1.2\overline{)72}$ **3.** $1.5\overline{)45}$

4. $7.5\overline{)-22.5}$ **5.** $4.8\overline{)16.8}$ **6.** $-2.7\overline{)11.07}$

Divide. Estimate to check whether your answer is reasonable.

7. $2.8\overline{)14}$ **8.** $-5.6\overline{)21}$ **9.** $3.2\overline{)48}$

10. $2.25\overline{)9}$ **11.** $2.4\overline{)6}$ **12.** $-1.25\overline{)65}$

13. Jessie used 2.7 gallons of gas to drive her car
72.9 miles. What was her car's gas mileage? ______________________

14. Ernesto bicycled 267 miles last week at an
average speed of 8.9 mi/h. How many hours
did he bicycle? ______________________

Holt Mathematics

Practice B
Dividing Decimals and Integers by Decimals

Divide.

1. $6 \div 0.25$

2. $78.74 \div 12.7$

3. $734.8 \div -1.67$

____

4. $56.525 \div 0.85$

5. $44.22 \div (-6.7)$

6. $- 6.46 \div 0.04$

____

Divide. Estimate to check whether your answer is reasonable.

7. $63 \div (-4.5)$

8. $8 \div 3.2$

9. $87 \div 7.25$

____

10. $- 36 \div 1.6$

11. $42 \div 4.8$

12. $90 \div 0.36$

____

13. Freddie used 6.75 gallons of gas to drive
155.25 miles. What was his car's gas mileage? _______________

14. The members of a book club met at a restaurant
for dinner. The total bill was $112.95 and they
shared the bill equally. Each person paid $12.55.
How many members are there in the book club? _______________

Holt Mathematics

<table><tr><td>LESSON
3-5</td><td>

Practice C
Dividing Decimals and Integers by Decimals
</td></tr></table>

Divide. Estimate to check whether each answer is reasonable.

1. $39.984 \div (-5.1)$ 　　　**2.** $355.94 \div 4.81$ 　　　**3.** $-96 \div 1.28$

__________　　　　　　　__________　　　　　　　__________

4. $34.146 \div 6.3$ 　　　　**5.** $21.384 \div (-2.4)$ 　　　**6.** $15 \div 0.4$

__________　　　　　　　__________　　　　　　　__________

7. $338.742 \div 0.54$ 　　　**8.** $401.598 \div (-0.81)$ 　　**9.** $166.17 \div (-0.29)$

__________　　　　　　　__________　　　　　　　__________

Simplify each expression.

10. $40 \cdot (1.8 \div 0.9) \cdot 0.25$ 　　　　　　**11.** $(32.4 \div 6.75) - 2.9 - 4.8$

__________　　　　　　　　　　__________

12. $4.4 + (2.7 - 6.5) \div 0.4 + 0.6$ 　　　**13.** $6.25 \cdot 3.6 \div 1.25 \div 3$

__________　　　　　　　　　　__________

14. $2.3 + (4.2 \cdot 1.32) \cdot 5 + 0.7$ 　　　**15.** $(-7.2 \cdot 0.9) + 3.9 + 6.48$

__________　　　　　　　　　　__________

16. The price of binders at the school store increases by
$0.25 each year. How long will it take for the price
to increase by $3.00?　　　　　　　　__________

　　　　37　　　　**Holt Mathematics**

LESSON 3-5 Reteach
Dividing Decimals and Integers by Decimals

To divide a decimal by a decimal:

Step 1: Make the divisor a whole number by moving the decimal point to the right.

Step 2: Move the decimal point in the dividend the same number of places. Remember to place the decimal in the quotient directly above the decimal point in the dividend.

Step 3: Divide.

Divide: $1.68 \div 0.3 \rightarrow 0.3\overline{)1.68} \rightarrow 3\overline{)16.8}$

$$\begin{array}{r} 5.6 \\ 3\overline{)16.8} \\ -15 \\ \hline 18 \\ -18 \\ \hline \end{array}$$

The divisor, 0.3, has 1 decimal place. Move the decimal point 1 place to the right in <u>both</u> the divisor and the dividend.

Complete.

1. $5.6\overline{)4.48}$

a. How many decimal places are in the divisor? ______

b. How many places do you need to move each decimal point? ______

c. Rewrite the division. ____________

d. Complete the division. What is the quotient? ______

Divide.

2. $5.2\overline{)3.64}$

3. $0.09\overline{)36.45}$

4. $0.59\overline{)0.708}$

Holt Mathematics

Reteach
3-5 Dividing Decimals and Integers by Decimals (continued)

Sometimes it is necessary to write zeros in the dividend.

$$6 \div 0.25 \longrightarrow 0.25\overline{)6} \longrightarrow 0.25\overline{)6.00} \longrightarrow 25\overline{)600}$$

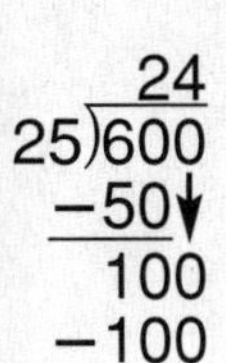

$$\begin{array}{r} 24 \\ 25\overline{)600} \\ -50 \\ \hline 100 \\ -100 \end{array}$$

The divisor, 0.25, has 2 decimal places. In order to move the decimal point 2 places to the right in the divisor and the dividend, you need to write 2 zeros in the dividend.

Complete.

5. $0.35\overline{)7}$

a. How many decimal places are in the divisor? _______

b. How many places do you need to move each decimal point? _______

c. How many zeros do you need to write in the dividend? _______

d. Complete the division. What is the quotient? _______

Divide.

6. $1.6\overline{)8}$ **7.** $0.12\overline{)19.2}$ **8.** $1.25\overline{)48}$

Holt Mathematics

LESSON 3-5

Challenge
Go For the Gold

A **golden rectangle** is a rectangle with a special relationship between the length and the width. The ancient Greeks believed that the shape of the golden rectangle was especially pleasing to the eye.

Discover the relationship between the length and width that makes a rectangle a golden rectangle.

1. The length of a golden rectangle is 5.26 centimeters and the width is 3.25 centimeters. Use a centimeter ruler to draw the rectangle.

2. The **golden ratio** is a value that characterizes the relationship between the length and width of a golden rectangle. Use the dimensions of the golden rectangle above. Divide the length by the width to find the golden ratio. Round to the nearest hundredth.

____

For each length given below, use the golden ratio to find the width to make a golden rectangle. Round your answers to the nearest hundredth.

3. length = 8 meters

4. length = 4.85 inches

5. length = 6.09 feet

6. length = 12 yards

7. length = 16.5 centimeters

8. length = 24.6 meters

9. length = 55 inches

10. length = 83.2 feet

11. length = 98.2 yards

12. length = 116 centimeters

13. length = 220.5 inches

14. length = 348.7 meters

Holt Mathematics

<table><tr><td>LESSON
3-5</td><td># Problem Solving
Dividing Decimals and Integers by Decimals</td></tr></table>

Write the correct answer.

1. Fran has a plank of wood 4.65 meters long. She wants to cut it into pieces 0.85 meters long. How many pieces of wood that length can she cut from the plank?

2. Jeremy has a box of 500 nails that weighs 1.35 kilograms. He uses 60 nails to build a birdhouse. How much do the nails in the birdhouse weigh?

3. Rhosanda is downloading a file from the Internet. The size of the file is 7.45 MB. The file is downloading at the rate of 0.095 MB per second. How many seconds will it take to download the entire file? Round your answer to the nearest second.

4. Sean has a piece of poster board with an area of 476.28 square centimeters. He cuts it into equal-sized squares, each with an area of 39.69 square centimeters. How many squares can he cut from the piece of poster board?

Choose the letter for the best answer.

5. Mount McKinley, in Alaska, is about 3.848 miles high. If a mountain climber can climb 0.25 miles per day, about how long, to the nearest day, would it take to climb Mount McKinley?

   A about 5 days

   B about 8 days

   C about 12 days

   D about 15 days

6. In 2000, a production worker in Japan who worked 38.5 hours in a week would have earned an average of $847. What was the hourly wage?

   F $2.20 per hour

   G $3.20 per hour

   H $22.00 per hour

   J $32.00 per hour

7. In the 2000 Summer Olympics, Michael Johnson ran the 400-meter race in 43.84 seconds. To the nearest hundredth, what was his speed in meters per second?

   A 9.12 meters per second

   B 10.96 meters per second

   C 1.09 meters per second

   D 91.24 meters per second

8. In the 2000 Summer Olympics, the United States relay team ran the 1,600-meter relay in 2 minutes, 56.35 seconds. To the nearest hundredth, what was the speed for the relay race in meters per second?

   F 2.83 meters per second

   G 6.24 meters per second

   H 9.07 meters per second

   J 28.39 meters per second

Holt Mathematics

Reading Strategies
LESSON 3-5 · *Use a Visual Model*

John has a piece of lumber 1.5 meters long. He needs to cut it into
pieces that are 0.3 meter long. How many pieces can he cut?
The number line shows a model of the problem.

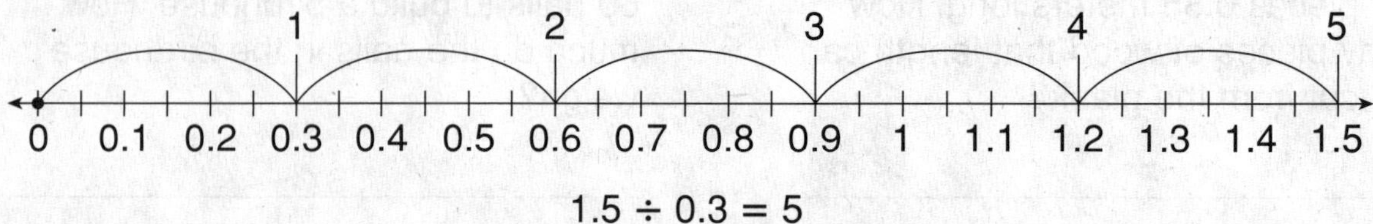

$$1.5 \div 0.3 = 5$$

Sarah has 15 feet of yarn. She needs to cut it into lengths of 3 feet
each. How many pieces can she cut? The number line shows a
model of the problem.

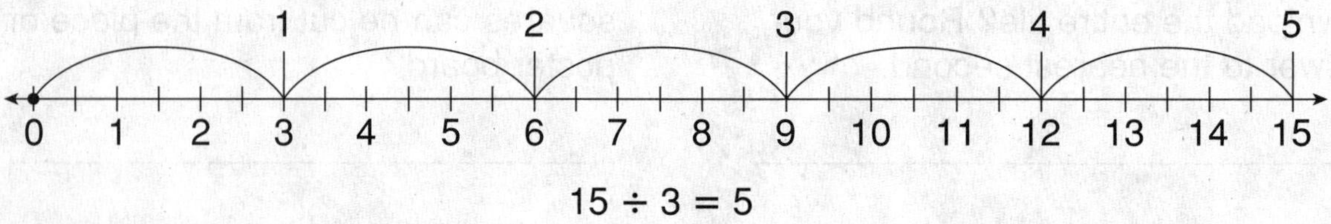

$$15 \div 3 = 5$$

Answer each question.

1. Compare the equations for the number lines above. What is the
   same about the equations?

2. What is different?

3. Compare the quotients of both problems. What do you notice?

4. How can you change 1.5 to 15?

5. How can you change 0.3 to 3?

6. If you moved the decimal point in **both** the divisor and the
   dividend, would the quotient change?

Holt Mathematics

Puzzles, Twisters & Teasers
By the Book!

Match the equations with their answers by writing the letter of
the correct answers on the lines. When you are finished, read
the letters from left to right and top to bottom as you would any
book. If your answers are correct, the letters will spell out the
title of a book. Write the book title on the blanks below.

32.1 **E**	0.9 **D**	−15 **N**	−5.3 **O**
5 **O**	6.25 **Y**	−0.5 **H**	120 **O**
0.4 **P**	309 **U**	0.046 **A**	14 **H**
5.6 **R**	−16 **E**	4 **A**	

3.78 ÷ 4.2 = ____   1.06 ÷ −0.2 = ____   7.5 ÷ 1.2 = ____

24 ÷ 0.2 = ____   9.27 ÷ 0.03 = ____   1.12 ÷ 0.08 = ____

−40 ÷ 2.5 = ____   5 ÷ 1.25 = ____   14 ÷ 2.5 = ____

1.15 ÷ 25 = ____   4.12 ÷ 10.3 = ____   −1.6 ÷ 3.2 = ____

12.0 ÷ 2.4 = ____   36 ÷ −2.4 = ____   96.3 ÷ 3 = ____

___ ___ ___   ___ ___ ___ ___

___ ___ ___ ___ ___ ___ ___ ?

**Now circle the name of someone who
might have been the author of this book.**

By:

Isabel Ringing   Justin Case

M. T. Wallet   Hugo First

Holt Mathematics

Practice A
Solving Equations Containing Decimals

Solve. Choose the letter for the best answer.

1. $t + 0.7 = 9$

   A $t = 9.7$ **C** $t = 6.3$

   B $t = 8.3$ **D** $t = 0.63$

2. $p - 1.6 = 11$

   F $p = 6.875$ **H** $p = 12.6$

   G $p = 9.4$ **J** $p = 17.6$

3. $\dfrac{h}{3} = 1.5$

   A $h = 0.5$ **C** $h = 9$

   B $h = 4.5$ **D** $h = 45$

4. $7z = 2.1$

   F $z = -4.9$ **H** $z = 14.7$

   G $z = 0.3$ **J** $z = 2.8$

Solve.

5. $x - 5.1 = 4.8$

6. $h + 6.9 = 12.7$

7. $k + 9.2 = -7.6$

8. $g - 4.44 = 2.4$

9. $0.18 + w = 0.75$

10. $m - 3.1 = 9.65$

11. $4.2n = 14.7$

12. $9.7j = 58.2$

13. $5.6p = -11.76$

14. $43.2 = 2.7y$

15. $64.6 = 6.8x$

16. $40.32 = 12.6m$

17. $\dfrac{s}{5.4} = 6$

18. $\dfrac{f}{0.8} = 7$

19. $\dfrac{d}{4.6} = 0.7$

20. $\dfrac{c}{0.4} = 1.75$

21. $\dfrac{h}{6.1} = 12$

22. $\dfrac{a}{0.35} = 8.4$

23. At the movie theater, a gift certificate for 15 tickets costs \$93.75. What is the cost of each ticket? _______________

24. Four couples are going to a concert. If each ticket costs \$5.90, how much will it cost to buy tickets for all 8 people? _______________

Holt Mathematics

Practice B
Solving Equations Containing Decimals

Solve.

1. $t + 0.77 = 9.3$

2. $p - 1.34 = -11.8$

3. $r + 2.14 = 7.8$

4. $3.65 + e = -1.4$

5. $w - 16.7 = 8.27$

6. $z - 17.2 = 7.13$

7. $p - 67.5 = 24.81$

8. $h + 26.9 = 12.74$

9. $k + 89.2 = -47.62$

10. $x - 0.45 = 5.97$

11. $1.08 + n = 15.72$

12. $y - 6.32 = 0.73$

13. $4.3p = 28.81$

14. $7.7j = 76.23$

15. $3.8g = -104.12$

16. $18.36 = 2.7y$

17. $99.96 = 6.8x$

18. $293.92 = 17.6c$

19. $\dfrac{e}{7.4} = 6.9$

20. $\dfrac{f}{12.7} = 15.6$

21. $\dfrac{d}{9.7} = 20.8$

22. $\dfrac{w}{-0.2} = 15.4$

23. $\dfrac{m}{9.8} = 1.7$

24. $\dfrac{s}{14.35} = -5.2$

25. Jeff paid a flat fee of \$269.50 for a year's worth of vet visits for his 4 cats. He made 14 visits during the year. What was the average cost per visit?

Holt Mathematics

LESSON 3-6 Practice C
Solving Equations Containing Decimals

Solve.

1. $t + 4.77 = 9.38$

2. $p - 1.34 = -17.08$

3. $2.55c = 19.89$

4. $3.65 + e = -10.74$

5. $16.17 - w = 8.27$

6. $34.932 = 4.92g$

7. $67.75 - a = 34.81$

8. $h + 26.99 = 11.74$

9. $k + 89.72 = -47.62$

10. $r + 2.14 = 27.84$

11. $6.45j = 36.765$

12. $2.91p = -24.444$

13. $29.406 = 7.54y$

14. $41.096 = 4.67x$

15. $17.52 - z = 4.13$

16. $\dfrac{a}{7.4} = 6.79$

17. $55.7 + b = 32.119$

18. $\dfrac{d}{9.7} = 20.78$

19. $0.378 + s = 2.918$

20. $v - 4.025 = 1.9024$

21. $\dfrac{f}{12.7} = 15.26$

22. $\dfrac{n}{0.38} = 5.89$

23. $\dfrac{t}{-1.88} = -6.2$

24. $\dfrac{w}{0.0056} = -4.7$

25. Sam lives 3.5 times farther from school than Sara does. Sam lives 13.3 miles from school. How far from school does Sara live? _______________

26. The Bow-Wow Pet Shop made a profit of $74,239 this year. This was $81,668 more than the profit from last year. What was the profit last year? _______________

Holt Mathematics

Reteach
3-6 Solving Equations Containing Decimals

You can solve equations with decimals the same way you solve equations with whole numbers. Remember to always perform the same calculation on both sides of the equation to keep the two sides equal.

- You can use addition to solve a subtraction equation involving decimals.

$$x - 1.45 = 6.7$$
$$x - 1.45 + \mathbf{1.45} = 6.7 + \mathbf{1.45}$$
$$x = 8.15$$

Addition undoes subtraction.

- You can use subtraction to solve an addition equation involving decimals.

$$n + 24.8 = -15.2$$
$$n + 24.8 - \mathbf{24.8} = -15.2 - \mathbf{24.8}$$
$$n = -40$$

Subtraction undoes addition.

Solve.

1. $e + 7.1 = 9.3$

   $e + 7.1 - 7.1 = 9.3 - 7.1$

   $e = $ _______

2. $x - 1.9 = 5.4$

   $x - 1.9 + $ _____ $= 5.4 + $ _____

   $x = $ _____

3. $w - 8.3 = -4.12$

   $w - 8.3$ _______ $= -4.12$ _______

   $w = $ _______

4. $b + 5.75 = -6.2$

   $b + 5.75$ _______ $= -6.2$ _______

   $b = $ _______

5. $t + 39.5 = 54.1$

6. $p - 29.4 = 3.7$

7. $r - 6.25 = -17.3$

8. $k + 9.8 = -11.9$

Holt Mathematics

LESSON 3-6 Reteach
Solving Equations Containing Decimals (continued)

- You can use division to solve a multiplication equation involving decimals.

$$3.6y = 9$$
$$\frac{3.6y}{3.6} = \frac{9}{3.6}$$
$$y = 2.5$$

Division undoes multiplication.

- You can use multiplication to solve a division equation involving decimals.

$$\frac{a}{4.2} = 18$$
$$4.2 \cdot \frac{a}{4.2} = 18 \cdot 4.2$$
$$a = 75.6$$

Multiplication undoes division.

Solve.

9. $$5.7g = 45.6$$

$$5.7g \div \underline{\hspace{1.5cm}} = 45.6 \div \underline{\hspace{1.5cm}}$$

$$g = \underline{\hspace{1.5cm}}$$

10. $$-6f = 8.04$$

$$-6f \underline{\hspace{0.6cm}} \underline{\hspace{0.6cm}} = 8.04 \underline{\hspace{0.6cm}} \underline{\hspace{0.6cm}}$$

$$f = \underline{\hspace{1.5cm}}$$

11. $$\frac{n}{0.14} = 15$$

$$\underline{\hspace{1.5cm}} \cdot \frac{n}{0.14} = 15 \cdot \underline{\hspace{1.5cm}}$$

$$n = \underline{\hspace{1.5cm}}$$

12. $$\frac{m}{6.3} = -9.1$$

$$\underline{\hspace{0.6cm}} \underline{\hspace{0.6cm}} \frac{m}{6.3} = -9.1 \underline{\hspace{0.6cm}} \underline{\hspace{0.6cm}}$$

$$m = \underline{\hspace{1.5cm}}$$

13. $$8y = 93.6$$

____

14. $$-3.4c = 20.74$$

____

15. $$\frac{s}{10.5} = 3.8$$

____

16. $$\frac{h}{0.4} = -7.2$$

____

Holt Mathematics

Challenge
The Pairs Event

When you solve a system of equations, you solve two different
equations at once. The solution to a system of equations is
an ordered pair (x, y).

Example: $2.5x = 50$
$x + y = 10.5$

First, solve the first equation for x: $x = 20$.

Then use the solution from the first equation to solve for y in the
second equation.

$x = 20$

$x + y = 10.5$
$20 + y = 10.5$
$y = -9.5$

The solution to the system of equations is the ordered
pair $(20, -9.5)$.

Solve each system of equations.

1. $x - 5.4 = 7.3$

$2x + y = 10$

____

2. $1.4x = 2.8$

$\dfrac{y}{x} = 9.6$

____

3. $\dfrac{x}{3.9} = 0.8$

$4y = x$

____

4. $x + 9.25 = 10.1$

$y - 3x = 1.7$

____

5. $6.4x = 2.56$

$x = 2y$

____

6. $x + 3.02 = 5$

$\dfrac{y}{3.5} = x$

____

7. $6x = -7.2$

$y - x = 3.4$

____

8. $\dfrac{x}{0.3} = -9$

$y + x = 0.6$

____

9. $x - 4.7 = -9$

$\dfrac{y}{x} = 0.5$

____

10. $x + 8.4 = -2$

$2x + y = 1.1$

____

11. $2.6x = 3.64$

$-3x = y$

____

12. $\dfrac{x}{0.4} = -3.2$

$x + y = 0.35$

____

Holt Mathematics

LESSON 3-6 Problem Solving
Solving Equations Containing Decimals

Write the correct answer.

1. The diameter of the secondary mirror in NASA's Hubble telescope is 0.3 meter. The primary mirror is 8 times as large. What is the diameter of the primary mirror?

2. A cubic centimeter of titanium weighs 4.507 grams. The same volume of gold weighs 19.3 grams. How much more does a cubic centimeter of gold weigh?

3. Brianna drives 3.35 miles to work every day. This is 1.75 miles less than Darius drives to work every day. How far does Darius drive to work?

4. The weight of an object on Mars is 0.38 times its weight on Earth. How much would a 125-pound person weigh on Mars?

Choose the letter for the best answer.

The table shows the orbital velocity of some of the planets.

5. How much greater is the orbital velocity of Mercury than that of Uranus?

   A 33.97 miles per second

   B 25.51 miles per second

   C 7.03 miles per second

   D 4.23 miles per second

6. How many miles does Mercury travel in an hour?

   F 1,784.4 miles

   G 107,064 miles

   H 10,706.4 miles

   J 17,844 miles

7. How many miles does Jupiter travel in a minute?

   A 8.12 miles

   B 81.2 miles

   C 48.72 miles

   D 487.2 miles

Planets' Orbital Velocity

Planet	Orbital Velocity (mi/s)
Mercury	29.74
Venus	21.76
Earth	18.5
Mars	14.99
Jupiter	8.12
Saturn	6.02
Uranus	4.23

8. During the time it takes Saturn to travel 32,508 miles, how much time has elapsed on Earth?

   F 5,400 minutes

   G 1,757.19 minutes

   H 195,698 seconds

   J 90 minutes

Holt Mathematics

LESSON 3-6 Reading Strategies
Compare and Contrast

Compare the steps for solving equations with whole numbers to the steps for solving equations with decimals.

Solving Equations with Whole Numbers	Example:
Step 1: This is a subtraction problem. Add to get x by itself.	$x - 145 = 1{,}720$
Step 2: Add **145** to both sides of the equation.	$x - 145 + 145 = 1{,}720$ $+ 145$
Step 3: Solve.	$x = 1{,}865$

Solving Equations with Decimals	Example:
Step 1: This is a subtraction problem. Add to get x by itself.	$x - 1.45 = 17.2$
Step 2: Add **1.45** to both sides of the equation.	$x - 1.45 + 1.45 = 17.2$ $+ 1.45$
Step 3: Solve.	$x = 18.65$

Use the chart to answer the following questions.

1. Compare the steps in solving an equation with whole numbers to the steps for an equation with decimals. What do you notice?

   ____

2. What is different about solving an equation with whole numbers and solving an equation with decimals?

   ____

   ____

Compare solving a multiplication equation with whole numbers to one with decimals: $3y = 702$; $3y = 7.02$. Answer each question.

3. What is the first step in solving both equations?

   ____

4. What operation will you use first in the two equations?

   ____

5. Compare the number you divide by on both sides of the whole number equation to the number you divide by in the decimal equation.

   ____

Holt Mathematics

Puzzles, Twisters & Teasers
LESSON 3-6
Messy Problems!

Solve the equations. Then use the variables to answer the riddle.

$-1.8 + g = -3.8$ $g =$ ________

$60t = 54$ $t =$ ________

$1.05 = -7n$ $n =$ ________

$h + 0.48 = 1.2$ $h =$ ________

$7.9 = i + 12.7$ $i =$ ________

$e + 0.81 = -6.3$ $e =$ ________

$\dfrac{l}{0.8} = 4.9$ $l =$ ________

$y - 4.1 = -5$ $y =$ ________

$1.2b = -1.44$ $b =$ ________

$\dfrac{a}{2.4} = 2.7$ $a =$ ________

$72 = 4.5r$ $r =$ ________

$3.2w = 8$ $w =$ ________

$\dfrac{d}{-6.4} = 0.85$ $d =$ ________

$s - 9.01 = 12.6$ $s =$ ________

$x + 30.34 = -22.87$ $x =$ ________

Why are basketball players sloppy eaters?

___ ___ ___ ___ ___ ___ ,
0.9   0.72   −7.11   −0.9   16   −7.11

___ ___ ___ ___ ___ ___
6.48   3.92   2.5   6.48   −0.9   21.61

___ ___ ___ ___ ___ ___ ___ ___ ___ .
−5.44   16   −4.8   −1.2   −1.2   3.92   −4.8   −0.15   −2

Holt Mathematics

LESSON 3-7 Practice A
Estimate with Fractions

Estimate each sum or difference.

1. $\dfrac{1}{6} + \dfrac{5}{8}$

$0 +$ _________ $=$ _________

2. $4\dfrac{7}{8} - 2\dfrac{1}{10}$

_________ $-$ _________ $=$ _________

3. $\dfrac{4}{5} + \dfrac{5}{9}$

4. $\dfrac{5}{12} + \dfrac{1}{5}$

5. $1\dfrac{7}{8} + 3\dfrac{5}{6}$

6. $\dfrac{7}{8} - \dfrac{2}{5}$

7. $\dfrac{8}{9} - \dfrac{5}{6}$

8. $8\dfrac{3}{5} - 3\dfrac{9}{10}$

Estimate each product or quotient.

9. $8\dfrac{1}{8} \cdot 9\dfrac{8}{9}$

$8 \cdot$ _________ $=$ _________

10. $39\dfrac{4}{5} \div 7\dfrac{9}{10}$

_________ $\div$ _________ $=$ _________

11. $3\dfrac{4}{5} \cdot 5\dfrac{9}{10}$

12. $5\dfrac{1}{5} \cdot 3\dfrac{5}{6}$

13. $6\dfrac{3}{8} \cdot 9\dfrac{1}{8}$

14. $23\dfrac{5}{8} \div 6\dfrac{1}{5}$

15. $29\dfrac{7}{9} \div 9\dfrac{3}{4}$

16. $9\dfrac{1}{5} \div 2\dfrac{7}{8}$

17. Erin hikes $3\dfrac{3}{4}$ miles in the morning and $5\dfrac{1}{8}$ miles in the afternoon. Estimate the number of miles Erin hikes in all.

____

18. A trail is $16\dfrac{3}{8}$ miles long. Max has hiked $4\dfrac{13}{16}$ miles along the trail so far. Estimate the number of miles Max has left to hike.

____

Holt Mathematics

Practice B
Estimate with Fractions

Estimate each sum or difference.

1. $\dfrac{5}{11} + \dfrac{4}{9}$

2. $\dfrac{6}{13} + \dfrac{8}{9}$

3. $\dfrac{9}{10} - \dfrac{4}{9}$

4. $1\dfrac{5}{8} - \dfrac{4}{7}$

5. $3\dfrac{7}{8} + \left(-\dfrac{2}{5}\right)$

6. $\dfrac{8}{9} - \dfrac{1}{12}$

7. $4\dfrac{5}{16} + 2\dfrac{9}{10}$

8. $11\dfrac{3}{7} - 5\dfrac{5}{6}$

9. $7\dfrac{1}{16} + \left(-\dfrac{11}{12}\right)$

Estimate each product or quotient.

10. $12\dfrac{2}{5} \div 5\dfrac{3}{4}$

11. $7\dfrac{7}{8} \cdot 4\dfrac{3}{5}$

12. $5\dfrac{1}{6} \cdot 3\dfrac{2}{9}$

13. $23\dfrac{7}{10} \div 4\dfrac{2}{5}$

14. $17\dfrac{11}{12} \div 8\dfrac{5}{9}$

15. $8\dfrac{7}{12} \cdot 6\dfrac{9}{10}$

16. $12\dfrac{3}{8} \cdot 6\dfrac{1}{6}$

17. $35\dfrac{2}{3} \div 3\dfrac{5}{7}$

18. $16\dfrac{5}{8} \cdot 2\dfrac{1}{5}$

19. A hallway has a length of $15\dfrac{3}{4}$ feet and a width of $4\dfrac{1}{12}$ feet. Estimate the area of the hallway in square feet.

20. A 6-week old puppy weighed $8\dfrac{7}{16}$ pounds. At 12 weeks of age, the same puppy weighed about $17\dfrac{3}{8}$ pounds. Estimate how much weight the puppy gained between the ages of 6 weeks and 12 weeks.

Holt Mathematics

<table><tr><td>**LESSON**
3-7</td><td></td></tr></table>

Practice C
Estimate with Fractions

Estimate each sum or difference.

1. $\dfrac{6}{11} + \dfrac{6}{7}$

2. $\dfrac{5}{12} - \dfrac{2}{5}$

3. $4\dfrac{4}{9} - 3\dfrac{1}{8}$

_______________   _______________   _______________

4. $5\dfrac{4}{7} + \dfrac{7}{12}$

5. $5\dfrac{7}{8} + \left(-2\dfrac{1}{16}\right)$

6. $6\dfrac{9}{10} - 2\dfrac{7}{16}$

_______________   _______________   _______________

7. $8\dfrac{11}{12} + 1\dfrac{2}{15} - 3\dfrac{6}{11}$

8. $3\dfrac{7}{15} + 5\dfrac{6}{7} - 1\dfrac{1}{8}$

9. $2\dfrac{3}{20} - 7\dfrac{7}{9} + 15\dfrac{9}{17}$

_______________   _______________   _______________

Estimate each product or quotient.

10. $2\dfrac{3}{5} \cdot 12\dfrac{1}{4}$

11. $5\dfrac{7}{9} \div 2\dfrac{3}{10}$

12. $20\dfrac{2}{9} \div \left(-4\dfrac{5}{7}\right)$

_______________   _______________   _______________

13. $7\dfrac{5}{16} \cdot \left(-3\dfrac{5}{8}\right)$

14. $16\dfrac{7}{9} \cdot 3\dfrac{9}{20}$

15. $14\dfrac{11}{15} \div 5\dfrac{2}{9}$

_______________   _______________   _______________

16. $-6\dfrac{5}{18} \cdot \left(-7\dfrac{11}{12}\right)$

17. $47\dfrac{13}{16} \div 3\dfrac{2}{11}$

18. $-31\dfrac{4}{5} \div 7\dfrac{8}{15}$

_______________   _______________   _______________

19. Carlos weighs $125\dfrac{1}{4}$ pounds. Hector weighs $17\dfrac{3}{4}$ pounds less than Carlos. About how much does Hector weigh?

____

20. A storage closet is $3\dfrac{3}{4}$ feet wide, $4\dfrac{1}{6}$ feet long, and $7\dfrac{11}{12}$ feet high. Estimate the volume of the closet.

____

Holt Mathematics

LESSON 3-7

Reteach

Estimate with Fractions

You can estimate sums and differences of fractions and mixed numbers by rounding the fractions or mixed numbers to the nearest $\frac{1}{2}$. Use a number line to help.

Estimate $\frac{3}{8} + \frac{9}{10}$.

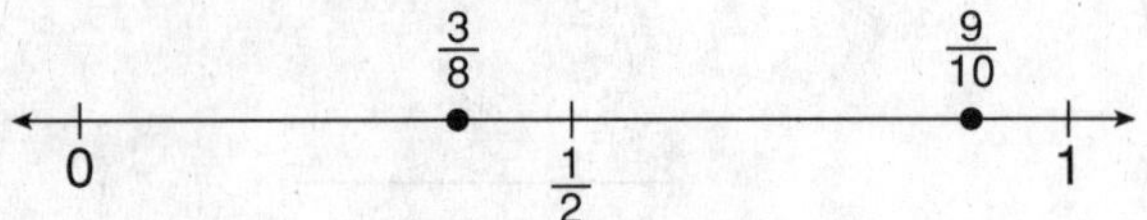

$\frac{3}{8}$ is about $\frac{1}{2}$. $\frac{9}{10}$ is about 1.

So, $\frac{3}{8} + \frac{9}{10}$ is about $\frac{1}{2} + 1$, or $1\frac{1}{2}$.

You can estimate products and quotients of mixed numbers by rounding the mixed numbers to the nearest whole numbers. Use a number line to help.

Estimate $2\frac{1}{4} \cdot 3\frac{5}{8}$.

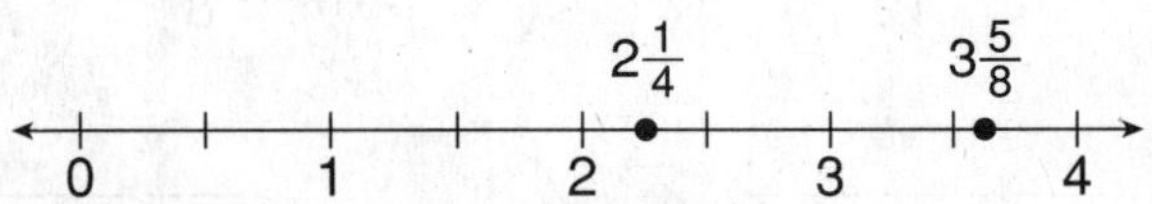

$2\frac{1}{4}$ is about 2. $3\frac{5}{8}$ is about 4.

So, $2\frac{1}{4} \cdot 3\frac{5}{8}$ is about $2 \cdot 4$, or 8.

Estimate.

1. $\frac{7}{8} - \frac{4}{7}$

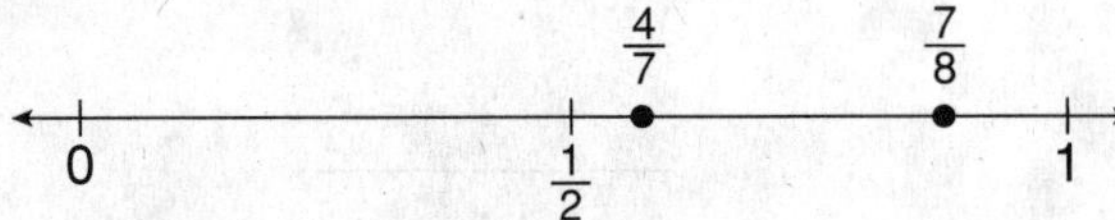

$\frac{7}{8}$ is about ____________

$\frac{4}{7}$ is about ____________.

$\frac{7}{8} - \frac{4}{7}$ is about ______ − ______,

or ______.

2. $4\frac{1}{6} \div 1\frac{5}{9}$

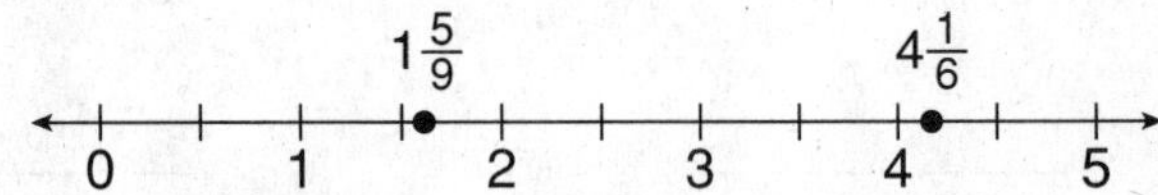

$4\frac{1}{6}$ is about ____________.

$1\frac{5}{9}$ is about ____________.

$4\frac{1}{6} \div 1\frac{5}{9}$ is about ______ ÷ ______,

or ______.

3. $\frac{11}{12} + \frac{2}{5}$

____

4. $6\frac{5}{8} - 4\frac{5}{6}$

____

5. $5\frac{8}{9} + 3\frac{7}{8}$

____

6. $3\frac{3}{8} \cdot 5\frac{7}{9}$

____

7. $23\frac{5}{6} \div 7\frac{7}{8}$

____

8. $5\frac{1}{4} \cdot 6\frac{2}{5}$

____

Holt Mathematics

Challenge

Don't Underestimate the Answer to a Riddle!

Circle the fractions whose estimated sums, differences products, or quotients equal the given solution. To find the answer to each riddle, write the letters below the circled fractions in order in the blanks.

Why did the student go to school in an airplane?

1. $\frac{7}{8} + \left[1\frac{6}{11} \text{ or } 4\frac{4}{5} \right] - \left[3\frac{7}{8} \text{ or } -\frac{9}{10} \right] \approx 3\frac{1}{2}$

    H   E   M   I

2. $5\frac{11}{12} + \left[6\frac{3}{8} \text{ or } -4\frac{5}{6} \right] \approx 1$

    P   G

3. $\left[\frac{6}{11} \text{ or } \frac{6}{7} \right] + \left[5\frac{9}{10} \text{ or } \frac{4}{19} \right] - \left[7\frac{2}{3} \text{ or } -4\frac{16}{19} \right] \approx 5\frac{1}{2}$

  H   T   Y   E   S   R

ANSWER: She wanted to get a _________________ education!

What kind of hair does the Atlantic Ocean have?

4. $3\frac{1}{10} + \left[\frac{9}{10} \text{ or } 4\frac{5}{12} \right] - \left[-6\frac{1}{20} \text{ or } -6\frac{9}{10} \right] \approx 11$

    W   H   U   A

5. $8\frac{4}{7} - \left[-6\frac{5}{9} \text{ or } -6\frac{1}{8} \right] \approx 15$

    V   R

6. $\left[2\frac{1}{3} \text{ or } 1\frac{6}{13} \right] - 4\frac{1}{10} \approx -2$

    Y   T

ANSWER: _________________

What can go across the country and still stay in its corner?

7. $\left[4\frac{5}{8} \text{ or } 6\frac{1}{8} \right] \cdot \left[8\frac{3}{10} \text{ or } 12\frac{9}{10} \right] \approx 48$

    T   S   T   R

8. $\left[20\frac{1}{5} \text{ or } 29\frac{7}{8} \right] \div \left[6\frac{3}{5} \text{ or } 4\frac{11}{13} \right] \approx 6$

    U   A   C   M

9. $63\frac{2}{5} \div \left[8\frac{4}{5} \text{ or } 6\frac{7}{9} \right] \approx 7$

    P   K

ANSWER: A _________________

57

Holt Mathematics

Problem Solving
Estimate with Fractions

Write the correct answer.

1. At the beach, Richard rides the waves on a boogie board that is $3\frac{2}{3}$ feet long. Laura rides a $7\frac{1}{2}$-foot surfboard. Estimate the difference in length of the 2 boards.

2. Jorgé had $5\frac{1}{3}$ jugs of apple cider. He used $2\frac{5}{6}$ jugs for a party. About how much apple cider does he have left?

3. Sari jogs $2\frac{3}{4}$ miles on Monday, $3\frac{5}{6}$ miles on Wednesday, and $2\frac{1}{3}$ miles on Friday. Estimate her total distance for the week.

4. Robert's hand is $2\frac{7}{8}$ inches wide. When Robert uses his hand to estimate the width of his desk, he finds that the desk is about $11\frac{3}{4}$ hands wide. About how many inches wide is the desk?

Choose the letter for the best answer.

This table shows the total amount of snow to fall in 5 cities during 2003.

Snowfall During 2003

City	Amount of Snow (in.)
Chicago, IL	$17\frac{2}{5}$
Indianapolis, IN	$44\frac{1}{10}$
Marquette, MI	$191\frac{4}{5}$
Moline, IL	$23\frac{4}{5}$
Providence, RI	$58\frac{9}{10}$

5. About how much snow all together fell in the two cities in Illinois?
   - **A** 40 inches
   - **B** 41 inches
   - **C** 44 inches
   - **D** 61 inches

6. About how much more snow fell in Providence than in Indianapolis?
   - **F** 35 inches
   - **G** 20 inches
   - **H** 15 inches
   - **J** 10 inches

7. Which city had about 11 times as much snow as Chicago?
   - **A** Indianapolis
   - **B** Marquette
   - **C** Moline
   - **D** Providence

8. About how much more snow fell in Indianapolis than in Moline?
   - **F** 20 inches
   - **G** 22 inches
   - **H** 24 inches
   - **J** 30 inches

Holt Mathematics

<table>
<tr><td>LESSON
3-7</td></tr>
</table>

Reading Strategies
Analyze Information

You can use some easy rules to help you estimate fractions.

· If the numerator is much smaller than the denominator, round to 0.	**Fractions Close to 0** $\frac{1}{5}$ $\frac{2}{25}$ $\frac{6}{50}$
· If the numerator is about half the value of the denominator, round to $\frac{1}{2}$.	**Fractions Close to $\frac{1}{2}$** $\frac{6}{13}$ $\frac{9}{20}$ $\frac{12}{23}$
· If the numerator and denominator are close to the same value, round to 1.	**Fractions Close to 1** $\frac{7}{8}$ $\frac{23}{25}$ $\frac{57}{60}$

This can help you estimate sums and differences of fractions.

$$\frac{5}{9} - \frac{14}{16}$$
$$\downarrow \qquad \downarrow$$
$$\frac{1}{2} - 1 = -\frac{1}{2}$$

$$3\frac{3}{8} + 1\frac{8}{9}$$
$$\downarrow \qquad \downarrow$$
$$3\frac{1}{2} + 2 = 5\frac{1}{2}$$

Write *close to 0, close to $\frac{1}{2}$*, or *close to 1* for each fraction.

1. $\frac{13}{25}$ _______________

2. $\frac{9}{75}$ _______________

3. $\frac{7}{9}$ _______________

4. $\frac{14}{16}$ _______________

5. $\frac{3}{20}$ _______________

6. $\frac{1}{9}$ _______________

7. $\frac{7}{15}$ _______________

Holt Mathematics

LESSON 3-7 — Puzzles, Twisters & Teasers
Math Makes Me Crabby!

Estimate the answers. Then solve the riddle.

H $35\frac{1}{3} \div 4\frac{9}{10}$ __________

T $2\frac{8}{9} + 1\frac{7}{8}$ __________

S $31\frac{7}{8} \div 3\frac{5}{9}$ __________

E $2\frac{2}{5} \cdot 2\frac{13}{15}$ __________

I $\frac{7}{9} - \frac{3}{8}$ __________

Y $\frac{3}{5} + \frac{6}{7}$ __________

F $4\frac{5}{7} \cdot 5\frac{1}{3}$ __________

A $8\frac{3}{4} + -6\frac{1}{5}$ __________

L $15\frac{1}{7} + 10\frac{8}{9}$ __________

R $\frac{5}{6} + \frac{2}{12}$ __________

Why are crabs so greedy?

___ ___ ___ ___   ___ ___ ___
5   7   6 $1\frac{1}{2}$ 3   1   6

___ ___ ___ ___ ___ ___ ___ ___ ___.
8   7   6   26   26   25 $\frac{1}{2}$ 8   7

Holt Mathematics

LESSON 3-8

Practice A
Adding and Subtracting Fractions

Add or subtract. Write each answer in simplest form.

1. $\frac{1}{2} + \frac{1}{2}$

2. $\frac{2}{5} + \frac{2}{5}$

3. $\frac{1}{4} + \frac{1}{4}$

4. $\frac{1}{8} + \frac{5}{8}$

5. $\frac{5}{6} + \frac{5}{6}$

6. $\frac{7}{20} + \frac{17}{20}$

7. $\frac{7}{12} - \frac{1}{12}$

8. $\frac{9}{16} - \frac{5}{16}$

9. $\frac{3}{10} - \frac{9}{10}$

10. $\frac{7}{8} - \frac{5}{8}$

11. $\frac{4}{9} - \frac{7}{9}$

12. $\frac{7}{15} - \frac{1}{15}$

13. $\frac{1}{3} + \frac{3}{4}$

14. $\frac{1}{4} + \frac{5}{6}$

15. $\frac{5}{8} + \frac{1}{6}$

16. $\frac{1}{4} - \frac{5}{6}$

17. $\frac{3}{4} - \frac{1}{5}$

18. $\frac{9}{10} - \frac{3}{5}$

19. Michael bought $\frac{1}{2}$ pound of Swiss cheese. He used $\frac{1}{4}$ pound for a sandwich. How much was left?

____

Holt Mathematics

Practice B
Adding and Subtracting Fractions

Add or subtract. Write each answer in simplest form.

1. $\dfrac{1}{5} + \dfrac{2}{5}$ 2. $\dfrac{4}{15} + \dfrac{8}{15}$ 3. $\dfrac{7}{12} - \dfrac{5}{12}$

_______________   _______________   _______________

4. $\dfrac{9}{10} - \dfrac{7}{10}$ 5. $\dfrac{7}{12} - \dfrac{11}{12}$ 6. $\dfrac{2}{7} + \dfrac{6}{7}$

_______________   _______________   _______________

7. $\dfrac{11}{15} + \dfrac{7}{15}$ 8. $\dfrac{3}{16} - \dfrac{1}{16}$ 9. $\dfrac{8}{21} + \dfrac{5}{21}$

_______________   _______________   _______________

10. $\dfrac{4}{5} - \dfrac{3}{4}$ 11. $\dfrac{3}{8} + \dfrac{1}{2}$ 12. $\dfrac{2}{5} - \dfrac{21}{25}$

_______________   _______________   _______________

13. $\dfrac{11}{12} + \dfrac{5}{6}$ 14. $\dfrac{7}{8} - \dfrac{5}{12}$ 15. $\dfrac{9}{10} + \dfrac{5}{6}$

_______________   _______________   _______________

16. $\dfrac{2}{5} - \dfrac{7}{8}$ 17. $\dfrac{5}{6} + \dfrac{11}{15}$ 18. $\dfrac{3}{4} - \dfrac{8}{15}$

_______________   _______________   _______________

19. The school track is $\dfrac{7}{8}$ mile in length. Sherri ran $\dfrac{2}{3}$ mile. How much farther does she have to go to get all the way around the track?

____

20. The Millers budget $\dfrac{1}{2}$ of their income for fixed expenses and $\dfrac{1}{8}$ of their income for savings. What fraction of their income is left?

____

Holt Mathematics

Practice C
Adding and Subtracting Fractions

Add or subtract. Write each answer in simplest form.

1. $\dfrac{7}{15} - \dfrac{4}{15}$

2. $\dfrac{7}{18} + \dfrac{11}{18}$

3. $\dfrac{6}{7} + \dfrac{8}{21}$

4. $\dfrac{2}{5} + \dfrac{7}{15}$

5. $\dfrac{5}{12} - \dfrac{4}{9}$

6. $\dfrac{7}{30} - \dfrac{9}{10}$

7. $\dfrac{8}{9} - \dfrac{5}{18}$

8. $\dfrac{7}{25} + \dfrac{4}{5}$

9. $\dfrac{3}{8} + \dfrac{5}{11}$

10. $\dfrac{1}{8} - \dfrac{19}{40}$

11. $\dfrac{5}{8} + \dfrac{7}{12}$

12. $\dfrac{5}{6} - \dfrac{5}{9}$

13. $\dfrac{7}{8} + \dfrac{4}{5} + \dfrac{9}{20}$

14. $\dfrac{11}{12} - \dfrac{5}{6} - \dfrac{2}{3}$

15. $-\dfrac{2}{3} + \dfrac{13}{15} - \dfrac{4}{5}$

16. $-\dfrac{7}{12} - \dfrac{1}{5} + \dfrac{3}{4}$

17. $\dfrac{7}{9} + \dfrac{2}{5} - \dfrac{4}{15}$

18. $\dfrac{13}{20} - \dfrac{2}{7} + \dfrac{11}{35}$

19. In 2001, the population of the United States was about $\dfrac{2}{7}$ billion people. It is projected that by 2025, there will be over $\dfrac{1}{3}$ billion people in the United States. How much will the population increase by 2025? _________________

20. Benjamin walks $\dfrac{11}{15}$ mile to work and then $\dfrac{3}{10}$ mile to the grocery store. How far does he walk? _________________

21. About $\dfrac{2}{5}$ of the population in the United States has blood type A. About $\dfrac{23}{50}$ have blood type O. What fraction of the population has either blood type A or O? _________________

Holt Mathematics

<table><tr><td>**LESSON**
3-8</td><td>

Reteach
Adding and Subtracting Fractions
</td></tr></table>

To add or subtract fractions with different denominators:

Step 1: Find the least common multiple of the denominators.

Step 2: Write both fractions with the least common multiple (LCM) as the denominator.

Step 3: Add or subtract the numerators, keeping the denominator the same. Write the answer in simplest form.

$$\frac{1}{4} + \frac{5}{6}$$

The LCM of 4 and 6 is 12.

$\frac{1}{4} = \frac{3}{12}$ and $\frac{5}{6} = \frac{10}{12}$

$\frac{1}{4} + \frac{5}{6} = \frac{3}{12} + \frac{10}{12} = \frac{3+10}{12} = \frac{13}{12} = 1\frac{1}{12}$

$$\frac{2}{3} - \frac{1}{2}$$

The LCM of 3 and 2 is 6.

$\frac{2}{3} = \frac{4}{6}$ and $\frac{1}{2} = \frac{3}{6}$

$\frac{2}{3} - \frac{1}{2} = \frac{4}{6} - \frac{3}{6} = \frac{4-3}{6} = \frac{1}{6}$

Add or subtract. Write each answer in simplest form.

1. $\frac{3}{5} + \frac{1}{3} = \frac{\ }{15} + \frac{\ }{15} = \frac{\ +\ }{15} = \frac{\ }{15}$

2. $\frac{8}{9} - \frac{1}{3} = \frac{8}{9} - \frac{\ }{9} = \frac{\ -\ }{9} = \frac{\ }{9}$

3. $\frac{2}{5} + \frac{1}{2} = \frac{\ }{10} + \frac{\ }{10} = \frac{\ +\ }{10} = \frac{\ }{10}$

4. $\frac{3}{4} - \frac{1}{3} = \frac{\ }{12} - \frac{\ }{12} = \frac{\ -\ }{12} = \frac{\ }{\ }$

5. $\frac{2}{3} + \frac{3}{4} = \frac{\ }{12} + \frac{\ }{12} = \frac{\ +\ }{12} = \frac{\ }{\ } = \frac{\ }{\ }$

6. $\frac{1}{4} - \frac{5}{8} = \frac{\ }{\ } - \frac{\ }{\ } = \frac{\ -\ }{\ } = \frac{\ }{\ }$

7. $\frac{1}{4} + \frac{2}{3}$

8. $\frac{9}{10} - \frac{1}{2}$

9. $\frac{7}{10} + \frac{2}{5}$

10. $\frac{11}{12} - \frac{2}{3}$

11. $\frac{1}{4} - \frac{7}{10}$

12. $\frac{3}{4} + \frac{4}{5}$

Holt Mathematics

LESSON 3-8

Challenge

Fraction Tic-Tac-Toe

Find each sum or difference. Write the answer in simplest terms. Cross off the answers in the tic-tac-toe board to win.

1. $\dfrac{3}{8} + \dfrac{1}{2} + \dfrac{3}{4} =$ _________

2. $-\dfrac{6}{7} + \dfrac{2}{3} + \left(-\dfrac{11}{21}\right) =$ _________

3. $\dfrac{8}{9} + \dfrac{1}{3} + \left(-\dfrac{7}{12}\right) =$ _________

4. $\dfrac{1}{6} + \left(-\dfrac{2}{3}\right) - \left(-\dfrac{5}{12}\right) =$ _________

5. $-\dfrac{7}{9} - \dfrac{5}{6} + \dfrac{1}{3} =$ _________

6. $\dfrac{7}{10} - \left(-\dfrac{1}{2}\right) + \left(-\dfrac{3}{5}\right) =$ _________

7. $\dfrac{9}{14} - \left(-\dfrac{2}{7}\right) - \left(-\dfrac{1}{2}\right) =$ _________

8. $-\dfrac{8}{15} - \dfrac{1}{3} + \dfrac{4}{5} =$ _________

9. $\dfrac{2}{9} - \dfrac{2}{5} + \left(-\dfrac{2}{15}\right) =$ _________

$\dfrac{1}{2}$	$\dfrac{1}{7}$	$\dfrac{23}{36}$
$\dfrac{2}{12}$	$-\dfrac{5}{7}$	$\dfrac{5}{7}$
$1\dfrac{5}{8}$	$-\dfrac{1}{6}$	$\dfrac{7}{8}$

$\dfrac{1}{12}$	1	$-\dfrac{3}{5}$
$\dfrac{5}{12}$	$1\dfrac{5}{18}$	$\dfrac{4}{5}$
$-\dfrac{1}{12}$	$-1\dfrac{5}{18}$	$\dfrac{3}{5}$

$-\dfrac{1}{7}$	$1\dfrac{3}{7}$	1
$\dfrac{1}{15}$	$-\dfrac{1}{15}$	$\dfrac{1}{7}$
$-\dfrac{2}{45}$	$-\dfrac{14}{45}$	$\dfrac{2}{45}$

Holt Mathematics

Problem Solving

LESSON 3-8

Adding and Subtracting Fractions

Write the correct answer.

1. During the recycling drive, $\frac{1}{5}$ of the material collected was bottles and $\frac{1}{4}$ was paper. Cardboard boxes made up $\frac{1}{10}$ of the material. How much of the total do these three items represent?

2. Decorations for school dances take $\frac{1}{5}$ of the student council's budget. Entertainment takes $\frac{3}{10}$ of the budget. What fraction of the budget is left?

3. The school environmental club made a poster to celebrate Earth Day. The poster is $\frac{7}{8}$ yard long and $\frac{2}{3}$ yard wide. What is the difference in the length and width of the poster?

4. Three students ran for president of the student council. Eddie received $\frac{1}{5}$ of the votes. Tamara received $\frac{3}{8}$ of the votes. Levi received the rest. Which student won the election?

Choose the letter for the best answer.

5. The Reeds budget $\frac{1}{3}$ of their income for rent and $\frac{1}{4}$ for food. How much of their budget is left?

   A $\frac{1}{2}$

   B $\frac{3}{4}$

   C $\frac{5}{12}$

   D $\frac{7}{12}$

6. Jasmine's CD collection is $\frac{3}{8}$ jazz, $\frac{1}{4}$ classical, and the rest rock music. What fraction of her CDs is rock music?

   F $\frac{1}{4}$

   G $\frac{3}{8}$

   H $\frac{1}{2}$

   J $\frac{5}{8}$

7. Wong has 2 boxes of saltwater taffy. One box contains $\frac{3}{4}$ pound, and the other box contains $\frac{7}{10}$ pound. How much taffy does he have all together?

   A $\frac{9}{10}$ pound

   B $1\frac{9}{20}$ pounds

   C $1\frac{1}{2}$ pounds

   D $1\frac{13}{20}$ pounds

8. In 1992, about $\frac{43}{100}$ people voted for Bill Clinton for President. About $\frac{1}{5}$ voted for Ross Perot and the rest voted for George Bush. About how many voted for George Bush?

   F about $\frac{37}{100}$

   G about $\frac{52}{100}$

   H about $\frac{57}{100}$

   J about $\frac{3}{5}$

Holt Mathematics

Reading Strategies

LESSON 3-8 *Use Fraction Strips*

It is easy to add and subtract fractions with like, or **common denominators.**

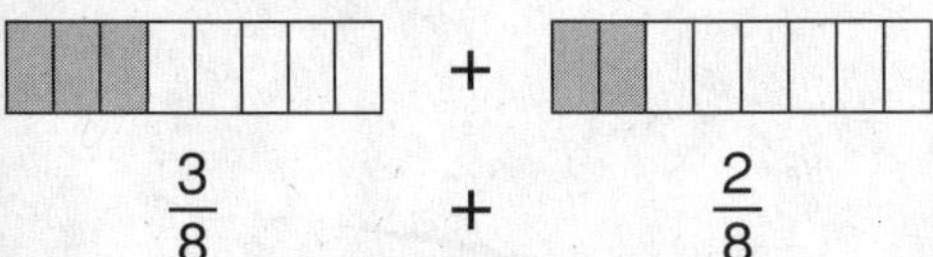

$$\frac{3}{8} \quad + \quad \frac{2}{8}$$

Use the fraction strips to answer each question.

1. What fractional part of the first fraction strip is shaded? ____________

2. What fractional part of the second fraction strip is shaded? ____________

3. Add the shaded parts of both fraction strips. What fractional part of both is shaded? ____________

4. When you add two fractions with common denominators, does the denominator change? ____________

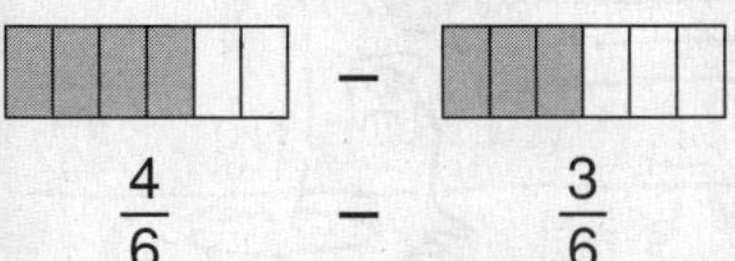

$$\frac{4}{6} \quad - \quad \frac{3}{6}$$

Use the fraction strips to answer each question.

5. What fractional part of the first fraction strip is shaded? ____________

6. What fractional part of the second fraction strip is shaded? ____________

7. When you subtract the second fraction strip from the first, what is the answer? ____________

8. When you subtract two fractions, what part of the fraction is subtracted, the numerator or the denominator? ____________

Holt Mathematics

LESSON 3-8   Puzzles, Twisters & Teasers
Jumping to Conclusions!

Solve the equations. Then answer the riddle.

D $\dfrac{1}{2} - \dfrac{3}{4}$ _______

B $\dfrac{5}{6} - \dfrac{1}{9}$ _______

E $\dfrac{1}{2} - \dfrac{7}{12}$ _______

C $\dfrac{4}{5} + \dfrac{6}{7}$ _______

O $\dfrac{5}{7} + \dfrac{1}{3}$ _______

A $\dfrac{3}{4} + \dfrac{2}{5}$ _______

L $\dfrac{1}{2} - \dfrac{2}{7}$ _______

U $\dfrac{2}{3} + \dfrac{1}{3}$ _______

S $\dfrac{7}{12} + \dfrac{6}{12}$ _______

T $\dfrac{21}{24} - \dfrac{1}{2}$ _______

I $\dfrac{1}{5} + \dfrac{2}{3}$ _______

H $\dfrac{5}{6} - \dfrac{1}{6}$ _______

W $\dfrac{3}{4} - \dfrac{11}{12}$ _______

N $\dfrac{5}{8} + \dfrac{7}{8}$ _______

If a cat can jump five feet in the air, then why can't it jump through a two-foot high window?

$$\frac{13}{18} \quad -\frac{1}{12} \quad 1\frac{23}{35} \quad 1\frac{3}{20} \quad 1 \quad 1\frac{1}{12} \quad -\frac{1}{12}$$

$$\frac{3}{8} \quad \frac{2}{3} \quad -\frac{1}{12} \quad\quad -\frac{1}{6} \quad \frac{13}{15} \quad 1\frac{1}{2} \quad -\frac{1}{4} \quad 1\frac{1}{21} \quad -\frac{1}{6}$$

$$\frac{13}{15} \quad 1\frac{1}{12} \quad\quad 1\frac{23}{35} \quad \frac{3}{14} \quad 1\frac{1}{21} \quad 1\frac{1}{12} \quad -\frac{1}{12} \quad -\frac{1}{4}$$

Holt Mathematics

Practice A
Adding and Subtracting Mixed Numbers

Add. Write each answer in simplest form.

1. $5\frac{1}{2} + 2\frac{1}{4}$

2. $3\frac{1}{4} + 4\frac{3}{4}$

3. $2\frac{1}{5} + 1\frac{2}{5}$

4. $1\frac{1}{2} + 3\frac{3}{8}$

5. $2\frac{6}{7} + 5\frac{4}{7}$

6. $3\frac{2}{9} + 3\frac{8}{9}$

7. $2\frac{5}{12} + 3\frac{1}{8}$

8. $2\frac{3}{4} + 5\frac{5}{6}$

9. $\frac{2}{3} + 2\frac{5}{8}$

Subtract. Write each answer in simplest form.

10. $4\frac{3}{4} - 2\frac{1}{4}$

11. $5\frac{5}{6} - 3\frac{1}{6}$

12. $7\frac{2}{3} - 4\frac{1}{3}$

13. $5\frac{1}{4} - 1\frac{3}{4}$

14. $6\frac{1}{3} - 5\frac{2}{3}$

15. $8\frac{1}{6} - 5\frac{5}{6}$

16. $9\frac{5}{6} - 6\frac{1}{4}$

17. $7\frac{3}{4} - 4\frac{5}{6}$

18. $8\frac{3}{8} - 4\frac{3}{4}$

19. Samson bicycled $8\frac{7}{8}$ miles on Friday and $5\frac{1}{4}$ miles on Saturday. How much farther did he ride on Friday?

Holt Mathematics

Practice B
Adding and Subtracting Mixed Numbers

Add. Write each answer in simplest form.

1. $7\frac{2}{7} + 6\frac{5}{7}$

2. $5\frac{4}{9} + 3\frac{7}{9}$

3. $4\frac{1}{3} + 8\frac{1}{4}$

____

4. $2\frac{7}{15} + 3\frac{11}{15}$

5. $6\frac{9}{10} + 1\frac{2}{5}$

6. $2\frac{3}{5} + 1\frac{11}{20}$

____

7. $5\frac{9}{10} + 2\frac{5}{8}$

8. $2\frac{11}{12} + 3\frac{7}{8}$

9. $1\frac{2}{3} + 5\frac{7}{9}$

____

Subtract. Write each answer in simplest form.

10. $7\frac{7}{9} - 3\frac{5}{9}$

11. $9\frac{7}{10} - 5\frac{3}{10}$

12. $4\frac{13}{15} - 1\frac{7}{15}$

____

13. $6\frac{2}{3} - 3\frac{3}{5}$

14. $10\frac{3}{4} - 6\frac{1}{3}$

15. $2\frac{3}{10} - 1\frac{7}{8}$

____

16. $8\frac{7}{12} - 6\frac{1}{3}$

17. $5\frac{7}{8} - 3\frac{9}{10}$

18. $7\frac{6}{7} - 6\frac{3}{4}$

____

19. Tucker ran $5\frac{3}{8}$ miles on Monday and $3\frac{3}{4}$ miles on Tuesday.
How far did he run on both days?

____

Holt Mathematics

Practice C
Adding and Subtracting Mixed Numbers

Add or subtract. Write each answer in simplest form.

1. $3\frac{11}{21} + 2\frac{8}{21}$

2. $2\frac{5}{26} + 5\frac{11}{13}$

3. $4\frac{2}{11} + 1\frac{6}{22}$

4. $15\frac{16}{19} - 8\frac{7}{19}$

5. $9\frac{13}{25} - 7\frac{2}{5}$

6. $4\frac{11}{20} - 3\frac{3}{10}$

7. $7\frac{3}{5} + 8\frac{1}{8}$

8. $4\frac{2}{3} + 6\frac{3}{7}$

9. $9\frac{5}{12} + 3\frac{3}{8}$

10. $8\frac{5}{6} - 5\frac{3}{5}$

11. $7\frac{7}{12} - 4\frac{11}{15}$

12. $10\frac{2}{3} - 5\frac{13}{18}$

13. $3\frac{2}{5} + 4\frac{1}{3} - 6\frac{7}{10}$

14. $1\frac{1}{4} + 6\frac{7}{8} + 5\frac{5}{16}$

15. $-4\frac{1}{2} - 9\frac{5}{6} - 5\frac{3}{4}$

16. $2\frac{1}{12} - 3\frac{2}{5} + 7\frac{4}{15}$

17. $1\frac{7}{8} + 2\frac{7}{20} - 1\frac{3}{10}$

18. $-6\frac{6}{7} - 3\frac{8}{35} - 1\frac{2}{5}$

Compare. Write <, >, or =.

19. $11\frac{1}{5} - 5\frac{4}{5} \;\square\; 7\frac{1}{3} - 1\frac{1}{2}$

20. $7\frac{2}{3} + 4\frac{1}{5} \;\square\; 7\frac{5}{6} + 4\frac{2}{7}$

21. $18\frac{1}{2} - 3\frac{3}{4} \;\square\; 7\frac{2}{3} + 7\frac{1}{12}$

22. $6\frac{1}{2} - 3\frac{1}{3} \;\square\; 5\frac{1}{3} - 2\frac{1}{5}$

23. The men's indoor pole vault record was set in 1993 at $20\frac{1}{6}$ feet. The women's record was set in 2001 at $15\frac{5}{12}$ feet. How much higher is the men's record than the women's record? _______________

24. Richard set a goal of running 10 miles per week. On Monday he ran $3\frac{3}{5}$ miles. On Friday he ran $3\frac{9}{10}$ miles. How much farther does he still have to run to meet his weekly goal? _______________

Holt Mathematics

Reteach
Adding and Subtracting Mixed Numbers

You can write mixed numbers as improper fractions before adding.

Add: $4\frac{5}{8} + 2\frac{7}{8}$

- Write improper fractions.

$$4\frac{5}{8} = \frac{37}{8} \text{ and } 2\frac{7}{8} = \frac{23}{8}$$

- The denominators are the same. Add and simplify.

$$\frac{37}{8} + \frac{23}{8} = \frac{60}{8} = 7\frac{4}{8} = 7\frac{1}{2}$$

Add: $1\frac{2}{5} + 3\frac{3}{4}$

- Write improper fractions.

$$1\frac{2}{5} = \frac{7}{5} \text{ and } 3\frac{3}{4} = \frac{15}{4}$$

- Find a common denominator. The LCD is 20.

$$\frac{7}{5} = \frac{28}{20} \text{ and } \frac{15}{4} = \frac{75}{20}$$

- Add and simplify.

$$\frac{28}{20} + \frac{75}{20} = \frac{103}{20} = 5\frac{3}{20}$$

Add. Write each answer in simplest form.

1. $4\frac{7}{10} + 2\frac{9}{10} = \frac{}{10} + \frac{}{10} = \frac{}{10} = \frac{}{10} = \underline{}$

2. $2\frac{1}{2} + 1\frac{3}{8} = \frac{}{2} + \frac{}{8} = \frac{}{8} + \frac{}{8} = \frac{}{8} = \frac{}{8}$

3. $3\frac{1}{5} + 2\frac{1}{3} = \frac{}{5} + \frac{}{3} = \frac{}{15} + \frac{}{15} = \frac{}{15} = \frac{}{15}$

4. $1\frac{2}{7} + 5\frac{3}{7}$

5. $5\frac{3}{8} + 3\frac{7}{8}$

6. $4\frac{4}{9} + 2\frac{2}{3}$

7. $2\frac{3}{5} + 3\frac{7}{10}$

8. $2\frac{3}{4} + 1\frac{5}{6}$

9. $4\frac{1}{3} + 2\frac{1}{2}$

Holt Mathematics

LESSON 3-9 **Reteach**

Adding and Subtracting Mixed Numbers (continued)

You can write whole numbers as fractions and mixed numbers as improper fractions before subtracting.

$6 - 2\frac{5}{9}$

- Write 6 as a fraction and $2\frac{5}{9}$ as an improper fraction.

  $6 = \frac{6}{1}$ and $2\frac{5}{9} = \frac{23}{9}$

- Find a common denominator. The LCD is 9.

  $\frac{6}{1} = \frac{54}{9}$

- Subtract and simplify.

  $\frac{54}{9} - \frac{23}{9} = \frac{31}{9} = 3\frac{4}{9}$

$4\frac{1}{4} - 1\frac{5}{6}$

- Write improper fractions.

  $4\frac{1}{4} = \frac{17}{4}$ and $1\frac{5}{6} = \frac{11}{6}$

- Find a common denominator. The LCD is 12.

  $\frac{17}{4} = \frac{51}{12}$ and $\frac{11}{6} = \frac{22}{12}$

- Subtract and simplify.

  $\frac{51}{12} - \frac{22}{12} = \frac{29}{12} = 2\frac{5}{12}$

Subtract. Write each answer in simplest form.

10. $6\frac{1}{4} - 1\frac{3}{4} = \frac{\quad}{4} - \frac{\quad}{4} = \frac{\quad}{4} = \frac{\quad}{4} = \quad$

11. $5 - 1\frac{2}{3} = \frac{\quad}{3} - \frac{5}{3} = \frac{\quad}{3} - \frac{\quad}{3} = \frac{\quad}{3} = \frac{\quad}{3}$

12. $5\frac{3}{4} - 4\frac{1}{8} = \frac{\quad}{4} - \frac{\quad}{8} = \frac{\quad}{8} - \frac{\quad}{8} = \frac{\quad}{8} = \frac{\quad}{8}$

13. $3\frac{1}{4} - 2\frac{5}{6} = \frac{\quad}{4} - \frac{\quad}{6} = \frac{\quad}{12} - \frac{\quad}{12} = \frac{\quad}{12}$

14. $8\frac{2}{5} - 6\frac{4}{5}$ **15.** $5 - 3\frac{4}{7}$ **16.** $10 - 5\frac{7}{10}$

_______________   _______________   _______________

17. $7\frac{1}{2} - 6\frac{3}{10}$ **18.** $2\frac{2}{3} - 1\frac{2}{5}$ **19.** $5\frac{1}{2} - 3\frac{2}{3}$

_______________   _______________   _______________

Holt Mathematics

LESSON 3-9 Challenge
Mixed Number Magic

**In a magic square, each row, column, and diagonal has the
same sum. Fill in the missing numbers to complete each magic
square below.**

Magic Square #1

$4\frac{1}{3}$		
$3\frac{1}{4}$		
$8\frac{2}{3}$	$1\frac{1}{12}$	

Magic Square #1 sum: _____________

Magic Square #2

$2\frac{1}{6}$	$4\frac{7}{8}$	
	$2\frac{17}{24}$	
		$3\frac{1}{4}$

Magic Square #2 sum: _____________

What is the difference between these two magic sums? _____________

Magic Square #3

6	$13\frac{1}{2}$	3
	$7\frac{1}{2}$	

Magic Square #3 sum: _____________

Magic Square #4

	$3\frac{3}{8}$	$\frac{3}{4}$
	$1\frac{7}{8}$	
	$\frac{3}{8}$	

Magic Square #4 sum: _____________

What is the difference between these two magic sums? _____________

Holt Mathematics

Problem Solving
Adding and Subtracting Mixed Numbers

Write the correct answer.

1. A female gray whale is $45\frac{1}{4}$ feet long. A male gray whale is $43\frac{1}{2}$ feet long. How much longer is the female than the male gray whale?

2. At birth, a pilot whale is $4\frac{3}{5}$ feet long. A newborn gray whale is $15\frac{1}{4}$ feet long. How much longer is the newborn gray whale?

3. A manatee weighs $\frac{1}{2}$ ton. A walrus weighs $1\frac{3}{4}$ tons. A narwhal weighs $1\frac{1}{2}$ tons. What is the total weight of all 3 animals?

4. A bottle-nosed dolphin can leap $15\frac{1}{8}$ feet out of the water. The world record high jump for a human is $8\frac{1}{24}$ feet. How much higher can a dolphin leap than a human?

Choose the letter for the best answer.

5. At a wildlife park, the killer whale show lasts $\frac{5}{8}$ of an hour. The guided tour of the park takes $2\frac{1}{2}$ hours. How long will it take to do both activities?

   A $3\frac{1}{8}$ hours

   B $2\frac{5}{8}$ hours

   C $3\frac{5}{8}$ hours

   D $1\frac{7}{8}$ hours

6. Jeremy walks $3\frac{1}{2}$ miles while visiting a wildlife park. Shawna hikes a $4\frac{3}{8}$-mile long nature trail. How much farther does Shawna walk?

   F $\frac{3}{8}$ of a mile

   G $\frac{5}{8}$ of a mile

   H $\frac{7}{8}$ of a mile

   J $1\frac{1}{8}$ of a mile

7. Dog food comes in $5\frac{7}{8}$-pound bags and $12\frac{3}{4}$-pound bags. Find the total weight of 2 small and 1 large bags.

   A $22\frac{7}{8}$ pounds

   B $23\frac{1}{2}$ pounds

   C $24\frac{1}{2}$ pounds

   D $25\frac{3}{4}$ pounds

8. A movie lasts $2\frac{1}{6}$ hours. A baseball game lasts $3\frac{2}{3}$ hours. How much longer does the game last?

   F $\frac{5}{6}$ hours

   G $1\frac{1}{6}$ hours

   H $1\frac{1}{3}$ hours

   J $1\frac{1}{2}$ hours

Holt Mathematics

<table><tr><td>LESSON
3-9</td><td>

Reading Strategies
Follow a Procedure
</td></tr></table>

To add mixed numbers that have fractions with common denominators, follow these steps.

$$2\frac{3}{7} + 1\frac{5}{7}$$

Step 1: Add fractions.

$$2\frac{3}{7}$$
$$+ 1\frac{5}{7}$$
$$\overline{\frac{8}{7}}$$

Step 2: Add integers.

$$2\frac{3}{7}$$
$$+ 1\frac{5}{7}$$
$$\overline{3}$$

Step 3: Simplify.

$$2\frac{3}{7}$$
$$+ 1\frac{5}{7}$$
$$\overline{3\frac{8}{7}} = 4\frac{1}{7}$$

Use the problem above to answer each question.

1. What part of the mixed numbers should you add first? ___________________

2. What part of the mixed numbers should you add next? ___________________

3. Do you need to simplify the sum? ___________________

When you subtract mixed numbers, you use the same steps as when you add them. However, when the step says add fractions or integers, you will subtract.

Use the problem $5\frac{3}{4} - 2\frac{1}{4}$ and the steps above to answer each question.

4. What is the first step in solving the problem?

____

5. What is the second step in solving the problem?

____

6. What is the answer of the problem?

76

Holt Mathematics

LESSON 3-9
Puzzles, Twisters & Teasers
Reaching the Right Solutions!

Decide whether or not each equation is correct. Circle the letters above your answers. Then answer the riddle.

1. $1\frac{2}{15} + 7\frac{1}{6} = 8\frac{3}{10}$

    T   B

  correct   incorrect

2. $1\frac{7}{9} - \frac{17}{18} = 1$

    Z   O

  correct   incorrect

3. $70\frac{4}{5} + 1\frac{4}{5} = 72\frac{3}{5}$

    R   A

  correct   incorrect

4. $7\frac{1}{3} + 8\frac{1}{5} = 16\frac{2}{5}$

    Y   E

  correct   incorrect

5. $6\frac{1}{4} + 8\frac{3}{4} = 14$

    X   A

  correct   incorrect

6. $3\frac{4}{5} + 3\frac{2}{5} = 7\frac{1}{5}$

    C   P

  correct   incorrect

7. $3\frac{1}{2} + 5\frac{1}{4} = 8\frac{3}{4}$

    H   V

  correct   incorrect

8. $14\frac{3}{5} - 8\frac{1}{2} = 7\frac{1}{5}$

    U   H

  correct   incorrect

9. $6\frac{3}{4} - 2\frac{3}{4} = 4$

    I   Q

  correct   incorrect

10. $9\frac{1}{6} - 4\frac{6}{9} = 5\frac{1}{3}$

    M   S

  correct   incorrect

11. $6\frac{1}{6} + 5\frac{3}{10} = 12\frac{1}{2}$

    L   H

  correct   incorrect

12. $8 - 2\frac{3}{4} = 5\frac{1}{4}$

    A   K

  correct   incorrect

13. $3\frac{1}{3} - 2\frac{5}{8} = 6\frac{7}{8}$

    J   N

  correct   incorrect

14. $1\frac{8}{9} + 4\frac{4}{9} = 5\frac{9}{9}$

    G   D

  correct   incorrect

15. $6\frac{2}{3} - 5\frac{1}{3} = 1\frac{1}{3}$

    S   F

  correct   incorrect

Why did the basketball player need long arms?

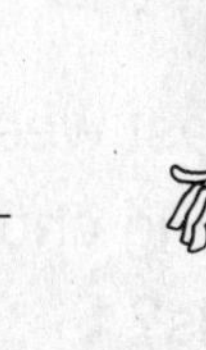

___ ___ ___   ___ ___ ___ ___ ___ ___

___ ___ ___ ___   ___ ___ ___ ___ ___ ___.

Holt Mathematics

LESSON 3-10 Practice A
Multiplying Fractions and Mixed Numbers

Multiply. Choose the letter for the best answer.

1. $7 \cdot \frac{1}{8}$

  A $\frac{1}{56}$ **C** $\frac{3}{4}$

  B $\frac{7}{15}$ **D** $\frac{7}{8}$

2. $\frac{2}{5} \cdot \frac{3}{4}$

  F $\frac{1}{4}$ **H** $\frac{3}{10}$

  G $\frac{2}{3}$ **J** $\frac{5}{9}$

3. $4 \cdot 3\frac{3}{5}$

  A $2\frac{2}{5}$ **C** $13\frac{1}{5}$

  B 12 **D** $14\frac{2}{5}$

4. $1\frac{1}{4} \cdot 2\frac{2}{3}$

  F $2\frac{1}{6}$ **H** $3\frac{2}{3}$

  G $3\frac{1}{3}$ **J** $3\frac{11}{12}$

Multiply. Write each answer in simplest form.

5. $4 \cdot \frac{1}{2}$

6. $8 \cdot \frac{1}{4}$

7. $10 \cdot \frac{1}{5}$

8. $\frac{1}{2} \cdot \frac{1}{4}$

9. $\frac{1}{4} \cdot -\frac{2}{3}$

10. $\frac{3}{4} \cdot \frac{2}{3}$

11. $-16 \cdot \frac{3}{4}$

12. $24 \cdot \frac{5}{6}$

13. $32 \cdot \frac{3}{8}$

14. $2\frac{1}{4} \cdot \frac{1}{2}$

15. $3\frac{1}{3} \cdot \frac{3}{5}$

16. $5\frac{1}{3} \cdot \frac{1}{4}$

17. $1\frac{1}{2} \cdot 1\frac{1}{5}$

18. $1\frac{2}{5} \cdot 2\frac{3}{4}$

19. $2\frac{2}{7} \cdot 3\frac{1}{8}$

20. Louis spent 12 hours last week practicing guitar. If $\frac{1}{4}$ of the time was spent practicing chords, how much time did he spend practicing chords? ______________

21. A banana bread recipe calls for $\frac{1}{4}$ tsp salt. Diane is making 5 loaves of banana bread. How much salt does she need? ______________

Holt Mathematics

Practice B
Multiplying Fractions and Mixed Numbers

Multiply. Write each answer in simplest form.

1. $5 \cdot \frac{1}{2}$

2. $9 \cdot \frac{3}{4}$

3. $6 \cdot -\frac{2}{5}$

____

4. $\frac{9}{15} \cdot \frac{5}{7}$

5. $\frac{9}{14} \cdot -\frac{7}{9}$

6. $\frac{7}{12} \cdot \frac{6}{14}$

____

7. $-12 \cdot \frac{3}{7}$

8. $15 \cdot \frac{5}{6}$

9. $21 \cdot \frac{3}{8}$

____

10. $2\frac{1}{3} \cdot \frac{3}{5}$

11. $3\frac{2}{5} \cdot \frac{1}{2}$

12. $4\frac{5}{6} \cdot \frac{2}{5}$

____

13. $2\frac{2}{5} \cdot \frac{2}{3}$

14. $3\frac{3}{4} \cdot \frac{2}{5}$

15. $8\frac{1}{6} \cdot \frac{3}{7}$

____

16. $2\frac{1}{3} \cdot 3\frac{3}{8}$

17. $1\frac{3}{5} \cdot 6\frac{2}{3}$

18. $2\frac{2}{5} \cdot 4\frac{5}{6}$

____

19. Rolf spent 15 hours last week practicing his saxophone. If $\frac{3}{10}$ of the time was spent practicing warm-up routines, how much time did he spend practicing warm-up routines?

____

20. A muffin recipe calls for $\frac{2}{5}$ tablespoon of vanilla extract for 6 muffins. Arthur is making 18 muffins. How much vanilla extract does he need?

____

Holt Mathematics

Practice C
Multiplying Fractions and Mixed Numbers

Multiply. Write each answer in simplest form.

1. $12 \cdot \dfrac{1}{7}$

____

2. $15 \cdot \dfrac{1}{4}$

____

3. $-7 \cdot \dfrac{1}{5}$

____

4. $\dfrac{8}{15} \cdot \left(-\dfrac{3}{4}\right)$

____

5. $\dfrac{6}{15} \cdot \dfrac{7}{18}$

____

6. $\dfrac{9}{11} \cdot \dfrac{22}{27}$

____

7. $\dfrac{2}{9} \cdot \dfrac{7}{10}$

____

8. $\dfrac{2}{15} \cdot \dfrac{5}{12}$

____

9. $\dfrac{7}{12} \cdot \dfrac{3}{7} \cdot \dfrac{1}{2}$

____

10. $3\dfrac{1}{2} \cdot 2\dfrac{5}{7}$

____

11. $3\dfrac{3}{8} \cdot 4\dfrac{2}{5}$

____

12. $-4\dfrac{1}{8} \cdot \dfrac{2}{9}$

____

13. $2\dfrac{2}{11} \cdot \dfrac{2}{3}$

____

14. $3\dfrac{3}{5} \cdot \dfrac{2}{9} \cdot \dfrac{2}{3}$

____

15. $5\dfrac{2}{5} \cdot 3\dfrac{1}{9}$

____

Complete each multiplication sentence.

16. $\dfrac{?}{5} \cdot \dfrac{3}{10} = \dfrac{3}{25}$

____

17. $\dfrac{1}{4} \cdot \dfrac{?}{7} = \dfrac{3}{14}$

____

18. $\dfrac{2}{3} \cdot \dfrac{?}{8} = \dfrac{5}{12}$

____

19. $\dfrac{?}{3} \cdot \dfrac{3}{8} = \dfrac{1}{4}$

____

20. $\dfrac{4}{5} \cdot \dfrac{?}{4} = \dfrac{3}{5}$

____

21. $\dfrac{3}{4} \cdot \dfrac{?}{7} = \dfrac{3}{7}$

____

22. A hippopotamus lives about $2\dfrac{4}{7}$ times as long as a tiger. A tiger lives an average of 16 years. How long does the average hippopotamus live?

____

23. A pie crust recipe calls for $\dfrac{4}{5}$ teaspoon of baking powder. Alice is making 13 pies. How much baking powder does she need?

____

Holt Mathematics

Reteach
Multiplying Fractions and Mixed Numbers

To multiply fractions and mixed numbers:
Step 1: Write any mixed numbers as improper fractions.
Step 2: Multiply the numerators.
Step 3: Multiply the denominators.
Step 4: Write the answer in simplest form.

Remember, positive times negative equals negative.

Multiply: $\dfrac{4}{9} \cdot \dfrac{3}{8}$

Divide numerator and denominator by 12, the GCF.

$$\dfrac{4}{9} \cdot \dfrac{3}{8} = \dfrac{4 \cdot 3}{9 \cdot 8}$$
$$= \dfrac{12}{72}$$
$$= \dfrac{1}{6}$$

Multiply: $6\dfrac{1}{4} \cdot \left(-1\dfrac{4}{5}\right)$

$$6\dfrac{1}{4} \cdot \left(-1\dfrac{4}{5}\right) = \dfrac{25}{4} \cdot \left(\dfrac{-9}{5}\right)$$
$$= \dfrac{25 \cdot (-9)}{4 \cdot 5}$$
$$= \dfrac{-225}{20}$$
$$= -11\dfrac{1}{4}$$

Multiply. Write each answer in simplest form.

1. $6 \cdot \dfrac{1}{9} = \dfrac{6 \cdot 1}{9} = \underline{} = \underline{}$

2. $-\dfrac{4}{5} \cdot \dfrac{5}{7} = -\dfrac{4 \cdot}{5 \cdot} = -\underline{} = -\underline{}$

3. $3\dfrac{1}{3} \cdot 9 = \dfrac{10}{3} \cdot 9 = \dfrac{10 \cdot}{} = \underline{} =$

4. $\dfrac{3}{10} \cdot 2\dfrac{1}{2} = \dfrac{3}{10} \cdot \dfrac{5}{2} = \dfrac{\cdot}{\cdot} = \underline{} = \underline{}$

5. $\dfrac{2}{7} \cdot \dfrac{7}{8}$

6. $-\dfrac{5}{9} \cdot \dfrac{3}{4}$

7. $\dfrac{9}{10} \cdot \left(-\dfrac{2}{3}\right)$

8. $2\dfrac{5}{8} \cdot \dfrac{2}{3}$

9. $\dfrac{1}{2} \cdot 4\dfrac{1}{4}$

10. $-\dfrac{2}{3} \cdot 1\dfrac{3}{4}$

11. $5\dfrac{1}{5} \cdot \left(-1\dfrac{2}{3}\right)$

12. $4\dfrac{1}{2} \cdot 1\dfrac{1}{9}$

13. $-2\dfrac{3}{4} \cdot \left(-1\dfrac{1}{3}\right)$

Holt Mathematics

LESSON 3-10 Challenge
Weigh Out in Space

How much do you think you would weigh on Mars? You can calculate anyone's weight on any of the planets by multiplying his or her weight times the planet's gravitational pull. Use the table to calculate what each animal listed below would weigh on different planets. Write your answers in simplest form.

Gravitational Pull of the Planets

Planet	Mercury	Venus	Mars	Jupiter	Saturn	Uranus	Neptune
Gravitational Pull	$\frac{19}{50}$	$\frac{9}{10}$	$\frac{19}{50}$	$2\frac{1}{2}$	$\frac{19}{20}$	$\frac{4}{5}$	$1\frac{1}{5}$

	Animal	Weight on Earth	Planet	Weight on Other Planet
1.	Mouse	$\frac{1}{16}$ pound	Neptune	
2.	Cat	$5\frac{1}{4}$ pounds	Mercury	
			Jupiter	
3.	Dog	$50\frac{1}{2}$ pounds	Saturn	
			Neptune	
4.	Gorilla	200 pounds	Mars	
			Uranus	
5.	Cow	1,000 pounds	Venus	
			Jupiter	
6.	Small parrot	$\frac{7}{8}$ pound	Mercury	
			Neptune	
7.	Macaw	$3\frac{1}{2}$ pounds	Neptune	
8.	Elephant	10,700 pounds	Uranus	

Holt Mathematics

Problem Solving

LESSON 3-10

Multiplying Fractions and Mixed Numbers

Write the correct answer.

1. Ariel's English homework is to read 24 pages. She reads $\frac{1}{8}$ of the assignment on the bus ride home. How many pages does she read on the bus?

2. When a group of 40 friends goes to the movies at a multiplex, $\frac{1}{5}$ of the group decides to watch a science fiction movie. How many of the group see the science fiction movie?

3. As of 1990, the American Indian population was about 2,000,000. About $\frac{1}{5}$ were Cherokee. About how many members of the Cherokee tribe were there in 1990?

4. Ron spends 3 hours painting a picture. Ashley spends $2\frac{2}{3}$ as long creating a sculpture. How long does Ashley work on her sculpture?

Choose the letter for the best answer.

5. One cup of dry dog food weighs $1\frac{4}{5}$ ounces. A K9 dog eats $6\frac{1}{3}$ cups of food a day. How many ounces of food does the dog eat each day?

A $4\frac{8}{15}$ ounces   **C** $11\frac{2}{5}$ ounces

B $8\frac{2}{15}$ ounces   **D** $3\frac{14}{27}$ ounces

6. Max has enough chicken to make $8\frac{1}{2}$ servings of salad. He needs $\frac{1}{8}$ pound of chicken per serving. How much chicken does he have?

F 8 pounds   **H** $8\frac{5}{8}$ pounds

G $1\frac{1}{16}$ pounds   **J** $2\frac{1}{2}$ pounds

7. A meteorite found in Willamette, Oregon, weighed $\frac{7}{10}$ as much as one found in Armanti, Western Mongolia. The meteorite found in Armanti weighed 22 tons. How much did the one in Oregon weigh?

A $31\frac{1}{5}$ tons

B $21\frac{3}{10}$ tons

C $22\frac{7}{10}$ tons

D $15\frac{2}{5}$ tons

8. Nicole takes part in a $12\frac{1}{2}$-mile walk-a-thon to raise money for charity. She stops $\frac{1}{2}$ of the way to rest. How much farther must Nicole walk to finish the walk-a-thon?

F $6\frac{1}{4}$ miles

G $6\frac{1}{2}$ miles

H 12 miles

J 25 miles

Holt Mathematics

LESSON 3-10 **Reading Strategies**
Use Fraction Strips

You can write a multiplication problem as a repeated addition problem.

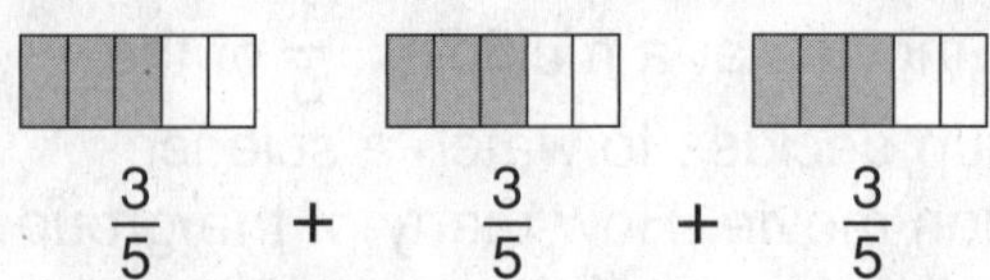

$$\frac{3}{5} \quad + \quad \frac{3}{5} \quad + \quad \frac{3}{5}$$

$\dfrac{3}{5} + \dfrac{3}{5} + \dfrac{3}{5}$ ← Repeated addition

three times three-fifths ← Multiplication

$3 \cdot \dfrac{3}{5}$

1. What fractional part of the fraction strips is shaded? _______________

2. How many fraction strips are there? _______________

3. Count the number of fractional parts that are shaded in all. How many are there? _______________

4. How can you find the answer to the problem above using addition?

____

You can also find the answer to the above problem using multiplication.

$3 \times \dfrac{3}{5} = \dfrac{9}{5}$

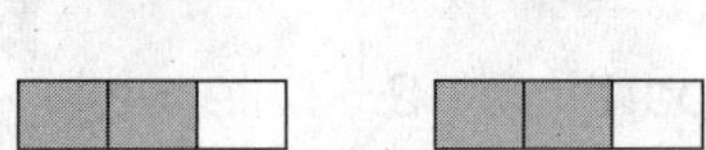

5. What fractional part of each fraction strip is shaded? _______________

6. How many of these fraction strips are there? _______________

7. Write a multiplication equation for this picture. _______________

Holt Mathematics

LESSON 3-10 · Puzzles, Twisters & Teasers
Wooden It Be Nice?

Decide whether or not each equation is correct. Circle the letter above your answer. Use the letters you circled to solve the riddle.

1. $2 \cdot \frac{3}{8} = 0.75$

I	L
correct	incorrect

2. $-15 \cdot \frac{2}{3} = -12$

M	T
correct	incorrect

3. $\frac{1}{4} \cdot \frac{4}{5} = \frac{1}{5}$

W	N
correct	incorrect

4. $\frac{1}{3} \cdot 4\frac{1}{2} = 1\frac{1}{2}$

O	P
correct	incorrect

5. $3\frac{3}{5} \cdot 1\frac{1}{12} = 3\frac{7}{10}$

Q	O
correct	incorrect

6. $\frac{1}{2} \cdot \frac{1}{3} \cdot \frac{1}{4} = \frac{1}{16}$

R	D
correct	incorrect

7. $\frac{1}{6} \cdot 12\frac{1}{2} = 24$

S	E
correct	incorrect

8. $2\frac{1}{4} \cdot 1\frac{1}{2} = 3\frac{3}{8}$

N	J
correct	incorrect

9. $2 \cdot 5\frac{2}{3} = 10\frac{1}{3}$

H	G
correct	incorrect

10. $75 \cdot 1\frac{1}{4} = 93.75$

O	K
correct	incorrect

What happened to the wooden plane with the wooden wheels and wooden engine?

___ _ ___

___ __ __ __ __ __ __ __ __

___ ___

Holt Mathematics

Practice A
Dividing Fractions and Mixed Numbers

Divide. Write each answer in simplest form.

1. $5 \div \dfrac{1}{2}$

2. $9 \div \dfrac{1}{3}$

3. $6 \div \dfrac{1}{4}$

____

4. $3 \div \dfrac{3}{4}$

5. $10 \div \dfrac{5}{6}$

6. $6 \div \dfrac{3}{8}$

____

Divide. Find each quotient in the box.

| $\dfrac{1}{5}$ | $\left(-\dfrac{1}{4}\right)$ | $\dfrac{1}{2}$ | $\left(-\dfrac{6}{11}\right)$ | $\dfrac{5}{7}$ | $\dfrac{7}{8}$ | 1 | $1\dfrac{1}{2}$ | 2 | $2\dfrac{6}{7}$ | 3 | 4 | $5\dfrac{1}{3}$ | $7\dfrac{1}{2}$ |

7. $\dfrac{9}{5} \div \dfrac{3}{5}$

8. $\dfrac{6}{7} \div \dfrac{3}{7}$

9. $\dfrac{1}{6} \div \dfrac{5}{6}$

____

10. $\dfrac{1}{3} \div \dfrac{2}{3}$

11. $\dfrac{3}{4} \div \dfrac{1}{2}$

12. $\dfrac{1}{6} \div \left(-\dfrac{2}{3}\right)$

____

13. $2\dfrac{2}{3} \div \dfrac{1}{2}$

14. $1\dfrac{1}{4} \div \dfrac{1}{6}$

15. $2\dfrac{1}{2} \div \dfrac{7}{8}$

____

16. $2\dfrac{1}{2} \div 3\dfrac{1}{2}$

17. $1\dfrac{1}{6} \div 1\dfrac{1}{3}$

18. $\left(-1\dfrac{1}{5}\right) \div 2\dfrac{1}{5}$

____

19. A jug holds 12 pints of juice. How many $1\dfrac{1}{2}$-pint bottles can be filled with that much juice?

____

Holt Mathematics

Practice B
Dividing Fractions and Mixed Numbers

Divide. Write each answer in simplest form.

1. $4 \div \frac{1}{2}$

2. $\frac{1}{5} \div \frac{1}{4}$

3. $\frac{1}{3} \div \frac{3}{5}$

____

4. $\frac{8}{9} \div \frac{2}{3}$

5. $-\frac{3}{8} \div \frac{3}{4}$

6. $\frac{7}{10} \div \frac{3}{5}$

____

7. $\frac{5}{12} \div \frac{2}{5}$

8. $\frac{3}{4} \div \frac{4}{9}$

9. $\frac{7}{12} \div \frac{3}{4}$

____

10. $-4\frac{1}{6} \div \frac{1}{3}$

11. $3\frac{1}{4} \div \frac{2}{5}$

12. $6\frac{1}{9} \div \frac{1}{6}$

____

13. $2\frac{1}{4} \div 1\frac{3}{4}$

14. $3\frac{3}{4} \div 2\frac{5}{6}$

15. $5\frac{1}{3} \div -1\frac{4}{5}$

____

16. $2\frac{1}{2} \div 2\frac{1}{3}$

17. $-1\frac{3}{4} \div 1\frac{1}{4}$

18. $7\frac{2}{3} \div 1\frac{1}{5}$

____

19. Burger Barn has $46\frac{2}{3}$ pounds of ground beef. How many $\frac{1}{3}$-pound burgers can be made using all the ground beef?

____

20. Roberto needs some roofing tiles to be cut from a large tile. How many tiles that are each $14\frac{3}{8}$ inches in length can he cut from a larger piece of tile that is $100\frac{5}{8}$ inches long?

____

Holt Mathematics

LESSON 3-11

Practice C
Dividing Fractions and Mixed Numbers

Divide. Write each answer in simplest form.

1. $17 \div \dfrac{4}{9}$

2. $23 \div \dfrac{5}{7}$

3. $39 \div \dfrac{7}{11}$

4. $-\dfrac{11}{15} \div \dfrac{7}{9}$

5. $\dfrac{9}{14} \div \dfrac{11}{21}$

6. $\dfrac{15}{19} \div \dfrac{3}{8}$

7. $\dfrac{3}{25} \div \dfrac{11}{15}$

8. $\dfrac{5}{39} \div -\dfrac{5}{13}$

9. $\dfrac{7}{17} \div \dfrac{14}{17}$

10. $5\dfrac{3}{7} \div \dfrac{4}{9}$

11. $7\dfrac{8}{9} \div \dfrac{5}{11}$

12. $-16\dfrac{2}{11} \div \dfrac{5}{6}$

13. $8\dfrac{5}{7} \div 3\dfrac{6}{7}$

14. $9\dfrac{3}{8} \div 2\dfrac{7}{8}$

15. $5\dfrac{3}{5} \div 1\dfrac{3}{8}$

16. $-\dfrac{2}{9} \div \dfrac{7}{8} \div \dfrac{1}{3}$

17. $\dfrac{3}{5} \div 4\dfrac{1}{4} \div \dfrac{2}{5}$

18. $3\dfrac{2}{3} \div \dfrac{2}{7} \div \dfrac{7}{9}$

19. Jorge can mow 1 lawn in $\dfrac{3}{4}$ hour. How many lawns can he mow in $3\dfrac{3}{4}$ hours?

20. How many $2\dfrac{1}{2}$-foot shelves can you cut from a $17\dfrac{1}{2}$-foot board?

Holt Mathematics

Reteach
Dividing Fractions and Mixed Numbers

Dividing fractions and mixed numbers is very much like multiplying fractions and mixed numbers. Just follow these steps:

> **Step 1:** Write any mixed numbers as improper fractions.
> **Step 2:** Invert the divisor.
> **Step 3:** Multiply and write the quotient in simplest form.

Divide: $1\frac{1}{8} \div \frac{1}{3}$

Step 1: $1\frac{1}{8} \div \frac{1}{3} = \frac{9}{8} \div \frac{1}{3}$

Step 2: $\frac{9}{8} \div \frac{1}{3} = \frac{9}{8} \cdot \frac{3}{1}$

Step 3: $\frac{9}{8} \cdot \frac{3}{1} = \frac{27}{8} = 3\frac{3}{8}$

Divide: $1\frac{1}{4} \div 3\frac{1}{3}$

Step 1: $1\frac{1}{4} \div 3\frac{1}{3} = \frac{5}{4} \div \frac{10}{3}$

Step 2: $\frac{5}{4} \div \frac{10}{3} = \frac{5}{4} \cdot \frac{3}{10}$

Step 3: $\frac{5}{4} \cdot \frac{3}{10} = \frac{15}{40} = \frac{3}{8}$

Divide. Write each answer in simplest form.

1. $\frac{4}{5} \div \frac{1}{2} = \frac{4}{5} \cdot \underline{} = \underline{} = \underline{}$

2. $\frac{5}{8} \div \frac{5}{6} = \frac{5}{8} \cdot \underline{} = \underline{} = \underline{}$

3. $2\frac{1}{2} \div 1\frac{3}{4} = \frac{}{2} \div \frac{}{4} = \frac{}{2} \cdot \underline{}$

$= \underline{} = \underline{} = \underline{}$

4. $2\frac{2}{3} \div 1\frac{1}{5} = \frac{}{3} \div \frac{}{5} = \frac{}{3} \cdot \underline{}$

$= \underline{} = \underline{} = \underline{}$

5. $\frac{3}{5} \div \frac{3}{10}$

6. $\frac{7}{8} \div \frac{1}{3}$

7. $\frac{5}{12} \div \frac{1}{2}$

8. $4\frac{1}{3} \div 1\frac{1}{9}$

9. $2\frac{1}{3} \div 1\frac{3}{4}$

10. $5\frac{5}{8} \div 2\frac{1}{2}$

Holt Mathematics

LESSON 3-11 Challenge
Animal Senior Citizens

The value of each division expression will show the average life expectancy of an animal. Complete the chart. Then use the information to fill in the blanks below.

	Animal	Division Expression	Life Expectancy (yr)
1.	Black bear	$2\frac{1}{2} \div \frac{5}{36}$	
2.	Chimpanzee	$40\frac{2}{5} \div 2\frac{1}{50}$	
3.	Dog	$15\frac{1}{2} \div 1\frac{7}{24}$	
4.	Hippopotamus	$4\frac{3}{8} \div \frac{7}{64}$	
5.	Mouse	$\frac{5}{6} \div \frac{5}{18}$	
6.	Guinea pig	$\frac{2}{3} \div \left(-\frac{3}{5}\right) \div \left(-\frac{5}{18}\right)$	
7.	Tiger	$-2\frac{1}{4} \div 3\frac{1}{2} \div \left(-\frac{9}{224}\right)$	
8.	Kangaroo	$\left(\frac{3}{4} + \frac{7}{8}\right) \div \frac{13}{56}$	

9. If you divide the average life span of a guinea pig by _______, you get the average life span of a hippopotamus.

10. If you divide the average life span of a kangaroo by $\frac{7}{20}$, you find the average life span of a horse, which is ___________.

11. A ________________ can live 37 years longer than a ________________. This is $5\frac{1}{2} \div 2\frac{3}{4}$ years longer than the life expectancy of an African elephant, which is ___________.

12. (Average life expectancy of a dog) $\div \frac{3}{8} \div 6\frac{2}{5}$ = (average life expectancy of a rabbit) = ___________.

Holt Mathematics

LESSON 3-11 Problem Solving
Dividing Fractions and Mixed Numbers

Write the correct answer.

1. The Wheeling Bridge in West Virginia is about $307\frac{4}{5}$ meters long. If you walk with a stride of about $\frac{3}{10}$ meter, how many steps would it take you to cross this suspension bridge?

2. The Flathead Rail Tunnel in Montana is about $7\frac{3}{4}$ miles long. If a train travels through the tunnel at a speed of about $1\frac{1}{2}$ miles per minute, how long will it take to pass from one end of the tunnel to the other?

3. A hiking trail is $6\frac{2}{3}$ miles long. It has 4 exercise stations, spaced evenly along the trail. What is the distance between each exercise station?

4. Jamal buys a strip of 25 postage stamps. The strip of stamps is $21\frac{7}{8}$ inches long. How long is each stamp?

Choose the letter for the best answer.

5. Matt wants to decorate his skateboard with decals. His skateboard is $28\frac{3}{4}$ inches long. The decals are $5\frac{1}{2}$ inches long. If Matt arranges them in a line from the front to the back, how many decals will fit?

   A 5 decals

   B 6 decals

   C 7 decals

   D 8 decals

6. A square floor tile measures $\frac{3}{4}$ square feet. How many tiles are required to cover a 200 square foot floor?

   F 150 tiles

   G 267 tiles

   H 275 tiles

   J 300 tiles

7. Bev buys a sleeve of ball bearings for her skateboard. Each of the bearings is $1\frac{1}{5}$ inches wide. The sleeve is $9\frac{3}{5}$ inches long. How many bearings are in the sleeve?

   A 5 bearings

   B 8 bearings

   C 9 bearings

   D 12 bearings

8. The average hamster weighs about $\frac{1}{4}$ pound. The total weight of all the hamsters in a cage in a pet store is $1\frac{1}{2}$ pounds. How many hamsters are in the cage?

   F 3 hamsters

   G 4 hamsters

   H 5 hamsters

   J 6 hamsters

Holt Mathematics

LESSON 3-11 **Reading Strategies**
Use a Visual Model

The Smith family has a two-and-a-half-foot-long sandwich to share.
One-half foot of the sandwich will serve one person. How many
one-half foot servings are in this sandwich?

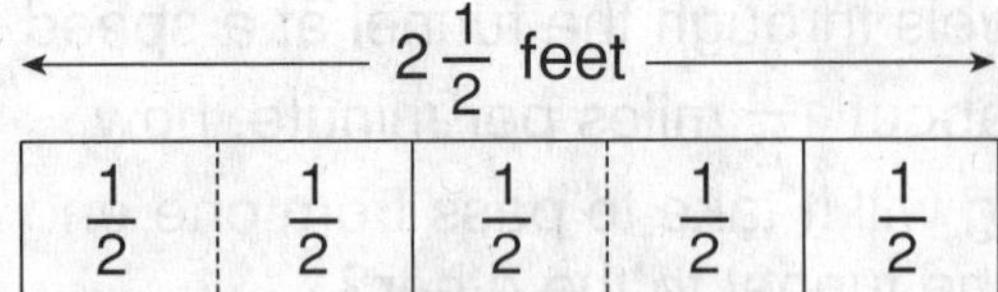

Use the model to answer each question.

1. How long is the sandwich?

____

2. How long is each serving?

____

3. If you divided the sandwich into $\frac{1}{2}$ ft servings, how many would
you have?

____

4. What is $2\frac{1}{2} \div \frac{1}{2}$?

____

Suppose you have two sandwiches.

| $\frac{1}{2}$ | $\frac{1}{2}$ | $\frac{1}{2}$ | $\frac{1}{2}$ | $\frac{1}{2}$ |

| $\frac{1}{2}$ | $\frac{1}{2}$ | $\frac{1}{2}$ | $\frac{1}{2}$ | $\frac{1}{2}$ |

5. How many feet are in both sandwiches?

____

6. What is $2\frac{1}{2} \times 2$?

____

7. Compare the answers to $2\frac{1}{2} \div \frac{1}{2}$ and $2\frac{1}{2} \times 2$. What do you notice?

____

Holt Mathematics

Puzzles, Twisters & Teasers
Stop Making Sense!

Decide whether or not each equation is correct. Circle the letter above your answer. Use the letters you circled to solve the riddle.

1. $\dfrac{2}{3} \div \dfrac{1}{5} = 3\dfrac{1}{3}$

    I   B

  correct   incorrect

2. $2 \div -\dfrac{7}{8} = 4\dfrac{1}{4}$

    D   T

  correct   incorrect

3. $\dfrac{3}{4} \div \dfrac{6}{7} = \dfrac{1}{3}$

    F   M

  correct   incorrect

4. $3\dfrac{3}{5} \div 9\dfrac{1}{7} = -3\dfrac{3}{5}$

    G   A

  correct   incorrect

5. $-4\dfrac{1}{3} \div 2\dfrac{1}{2} = -1\dfrac{11}{15}$

    K   H

  correct   incorrect

6. $\dfrac{3}{5} \div 6 = \dfrac{1}{10}$

    E   J

  correct   incorrect

7. $9 \div 1\dfrac{1}{2} = 6$

    S   L

  correct   incorrect

8. $10 \div \dfrac{5}{9} = \dfrac{3}{10}$

    O   C

  correct   incorrect

9. $-4\dfrac{4}{5} \div \dfrac{6}{7} = -5\dfrac{3}{5}$

    E   P

  correct   incorrect

10. $\dfrac{2}{3} \div \dfrac{8}{9} = 1$

    Q   N

  correct   incorrect

11. $\dfrac{3}{4} \div \dfrac{6}{7} = \dfrac{6}{8}$

    R   T

  correct   incorrect

12. $\dfrac{4}{9} \div 12 = 1\dfrac{3}{4}$

    V   S

  correct   incorrect

What did the little boy say when he learned how to count money?

__ ___

__ ___ _____

__ ___ ___ ___ ___!

Holt Mathematics

LESSON 3-12 Practice A
Solving Equations Containing Fractions

Solve. Choose the letter for the best answer.

1. $t - \dfrac{3}{4} = \dfrac{1}{4}$

   A $t = \dfrac{1}{4}$ **C** $t = \dfrac{3}{4}$

   B $t = \dfrac{1}{2}$ **D** $t = 1$

2. $g - \dfrac{3}{8} = \dfrac{1}{8}$

   F $g = \dfrac{1}{4}$ **H** $g = \dfrac{3}{4}$

   G $g = \dfrac{1}{2}$ **J** $g = 1$

3. $k + \dfrac{7}{12} = \dfrac{11}{12}$

   A $k = \dfrac{1}{4}$ **C** $k = \dfrac{1}{2}$

   B $k = \dfrac{1}{3}$ **D** $k = 1$

4. $n + \dfrac{2}{5} = \dfrac{4}{5}$

   F $n = \dfrac{2}{5}$ **H** $n = \dfrac{4}{5}$

   G $n = \dfrac{3}{5}$ **J** $n = 1$

5. $f + \dfrac{1}{6} = \dfrac{5}{6}$

   A $f = \dfrac{1}{6}$ **C** $f = \dfrac{1}{2}$

   B $f = \dfrac{1}{3}$ **D** $f = \dfrac{2}{3}$

6. $\dfrac{1}{4}s = 4$

   F $s = \dfrac{1}{2}$ **H** $s = 4$

   G $s = 1$ **J** $s = 16$

7. $\dfrac{2}{5}a = 10$

   A $a = 5$ **C** $a = 25$

   B $a = 10$ **D** $a = 50$

8. $\dfrac{1}{6}t = \dfrac{1}{2}$

   F $t = 1$ **H** $t = 6$

   G $t = 3$ **J** $t = 12$

Solve. Write each answer in simplest form.

9. $p - \dfrac{1}{4} = \dfrac{1}{6}$

10. $d - \dfrac{2}{5} = \dfrac{3}{10}$

11. $y + \dfrac{5}{8} = \dfrac{3}{4}$

____

12. $\dfrac{3}{4}m = \dfrac{5}{6}$

13. $\dfrac{1}{2}x = \dfrac{5}{8}$

14. $\dfrac{5}{6}r = \dfrac{3}{10}$

____

15. Eunice walked $\dfrac{1}{4}$ mile from home to the store on her way to a friend's house. If the store is $\dfrac{1}{3}$ of the way to her friend's house, how far is her friend's house from home?

____

**Holt Mathematics

Practice B
Solving Equations Containing Fractions

Solve. Write each answer in simplest form.

1. $t - \dfrac{3}{7} = \dfrac{4}{7}$

2. $g - \dfrac{5}{16} = \dfrac{3}{16}$

3. $k - \dfrac{3}{10} = \dfrac{2}{5}$

____

____

____

4. $n + \dfrac{1}{7} = \dfrac{2}{3}$

5. $j + \dfrac{5}{6} = \dfrac{17}{18}$

6. $f + \dfrac{5}{12} = \dfrac{3}{4}$

____

____

____

7. $\dfrac{1}{4}s = \dfrac{3}{4}$

8. $\dfrac{1}{5}a = \dfrac{1}{2}$

9. $\dfrac{4}{5}h = \dfrac{8}{9}$

____

____

____

10. $p - \dfrac{2}{3} = \dfrac{5}{8}$

11. $d - \dfrac{3}{5} = \dfrac{7}{10}$

12. $y - \dfrac{2}{7} = 3\dfrac{1}{4}$

____

____

____

13. $c + \dfrac{5}{12} = 2\dfrac{1}{6}$

14. $w + \dfrac{4}{15} = 3\dfrac{1}{3}$

15. $z + \dfrac{6}{7} = 2\dfrac{3}{5}$

____

____

____

16. $\dfrac{5}{6}m = \dfrac{8}{9}$

17. $\dfrac{1}{2}x = 3\dfrac{7}{15}$

18. $\dfrac{1}{5}r = 2\dfrac{2}{3}$

____

____

____

19. Sarabeth ran $1\dfrac{2}{5}$ miles on a path around the park. This was $\dfrac{5}{8}$ of the distance around the park. What is the distance around the park?

____

20. An interior decorator bought $12\dfrac{1}{2}$ yards of material to make drapes. He used $8\dfrac{2}{3}$ yards on 1 pair of drapes. How much material does he have left?

____

Holt Mathematics

Practice C
Solving Equations Containing Fractions

Solve. Write each answer in simplest form.

1. $t - \dfrac{12}{19} = \dfrac{5}{19}$

2. $g - \dfrac{5}{22} = \dfrac{3}{11}$

3. $k - \dfrac{3}{4} = \dfrac{2}{3}$

4. $n + \dfrac{8}{15} = \dfrac{3}{10}$

5. $j + \dfrac{11}{12} = \dfrac{13}{18}$

6. $f + \dfrac{8}{11} = \dfrac{4}{5}$

7. $\dfrac{1}{7}s = 1\dfrac{1}{4}$

8. $\dfrac{3}{5}a = 3\dfrac{2}{7}$

9. $-\dfrac{4}{9}h = -6\dfrac{2}{7}$

10. $p - \dfrac{2}{7} = 2\dfrac{3}{8}$

11. $d + \dfrac{5}{12} = -2\dfrac{2}{15}$

12. $y - \dfrac{2}{5} = 4\dfrac{4}{9}$

13. $-\dfrac{7}{10} + c = 5\dfrac{3}{20}$

14. $w + 1\dfrac{7}{11} = 4\dfrac{2}{5}$

15. $z + 3\dfrac{1}{6} = 6\dfrac{2}{9}$

16. $\dfrac{5}{8}m = 2\dfrac{4}{7}$

17. $-\dfrac{1}{11}x = 2\dfrac{1}{10}$

18. $\dfrac{3}{7}r = 3\dfrac{1}{3}$

19. A grizzly bear can run $1\dfrac{1}{5}$ times as fast as an elephant. A grizzly bear can run 30 mi/h. How fast can an elephant run?

20. A fruit salad recipe with bananas, strawberries, and grapes requires $3\dfrac{1}{2}$ pounds of fruit. You need $1\dfrac{3}{4}$ pounds of bananas and $\dfrac{7}{8}$ of a pound of strawberries. How many pounds of grapes are needed?

Holt Mathematics

Reteach
Solving Equations Containing Fractions

You can use addition to solve a subtraction equation involving fractions.

$$x - \frac{4}{9} = \frac{1}{3}$$
$$x - \frac{4}{9} + \frac{4}{9} = \frac{1}{3} + \frac{4}{9}$$
$$x = \frac{3}{9} + \frac{4}{9}$$
$$x = \frac{7}{9}$$

Remember, addition undoes subtraction.

You can use subtraction to solve an addition equation involving fractions.

$$n + \frac{2}{5} = \frac{9}{10}$$
$$n + \frac{2}{5} - \frac{2}{5} = \frac{9}{10} - \frac{2}{5}$$
$$n = \frac{9}{10} - \frac{4}{10}$$
$$n = \frac{5}{10} = \frac{1}{2}$$

Remember, subtraction undoes addition.

Solve. Write each answer in simplest form.

1. $d - \frac{1}{6} = \frac{3}{4}$

$$d - \frac{1}{6} + \text{—} = \frac{3}{4} + \text{—}$$
$$d = \frac{\ }{12} + \frac{\ }{12}$$
$$d = \frac{\ }{12}$$

2. $y + \frac{4}{5} = \frac{14}{15}$

$$y + \frac{4}{5} - \text{—} = \frac{14}{15} - \text{—}$$
$$y = \frac{14}{15} - \frac{\ }{15}$$
$$y = \frac{\ }{15}$$

3. $t - \frac{1}{8} = \frac{3}{4}$

4. $k + \frac{1}{2} = 1\frac{5}{8}$

5. $a - \frac{3}{5} = \frac{7}{10}$

_____________   _____________   _____________

Holt Mathematics

Reteach
Solving Equations Containing Fractions (continued)

You can use division to solve a multiplication equation involving fractions. Multiply both sides of the equation by the reciprocal of the coefficient of the variable.

$$3y = \frac{9}{10}$$

$$3y \cdot \frac{1}{3} = \frac{9}{10} \cdot \frac{1}{3}$$ ∘ ○ ○ ○ ○   The reciprocal of 3 is $\frac{1}{3}$.

$$y = \frac{9}{10} \cdot \frac{1}{3}$$

$$y = \frac{9}{30} = \frac{3}{10}$$

$$\frac{5}{6}a = \frac{1}{2}$$

$$\frac{5}{6}a \cdot \frac{6}{5} = \frac{1}{2} \cdot \frac{6}{5}$$ ∘ ○ ○ ○ ○   The reciprocal of $\frac{5}{6}$ is $\frac{6}{5}$.

$$a = \frac{1}{2} \cdot \frac{6}{5}$$

$$a = \frac{6}{10} = \frac{3}{5}$$

Solve. Write each answer in simplest form.

6. $8x = 3\frac{1}{5}$

$$8x = \frac{}{5}$$

$$8x \cdot \frac{}{} = \frac{}{5} \cdot \frac{}{}$$

$$x = \frac{}{5} \cdot \frac{}{}$$

$$x = \frac{}{} = \frac{}{}$$

7. $\frac{2}{3}k = \frac{5}{6}$

$$\frac{2}{3}k \cdot \frac{}{} = \frac{5}{6} \cdot \frac{}{}$$

$$k = \frac{5}{6} \cdot \frac{}{}$$

$$k = \frac{}{} = \frac{}{} = \frac{}{}$$

8. $\frac{3}{4}d = 5$

9. $6y = \frac{2}{3}$

10. $\frac{1}{5}s = \frac{5}{8}$

_________   _________   _________

Holt Mathematics

Challenge
Shopping List Equation

On the right is the list of ingredients for a recipe for bean and cheese tacos. The recipe serves 4 people. You are making bean and cheese tacos for 18 people.

Use the ingredients to answer the questions and solve the equations.

Bean and Cheese Tacos
$\frac{1}{2}$ pound kidney beans
1 clove garlic, chopped
4 flour tortillas
1 cup ricotta cheese
$\frac{1}{4}$ cup parmesan cheese
$\frac{1}{4}$ cup green onions, chopped
Serves 4

1. By what factor do you need to multiply each ingredient to serve 18 people? Solve the equation to find out. $4x = 18$

   $x =$ _______________________

2. How many pounds of kidney beans do you need to serve 18 people?

   ____

3. Kidney beans come in $\frac{3}{4}$-pound cans. How many cans do you need to serve 18 people? Write and solve an equation.

   ____

4. Ricotta cheese is on sale in $1\frac{1}{4}$-cup tubs. How many tubs do you need to serve 18 people? Write and solve an equation.

   ____

5. You need $1\frac{1}{8}$ cups of Parmesan cheese. You have $\frac{1}{4}$ cup. Write and solve an equation to find how much more you need.

   ____

6. There are 4 green onions in $\frac{1}{4}$ cup. Write and solve an equation to find how many green onions you need.

   ____

7. You have 8 cloves of garlic and you need $4\frac{1}{2}$. How many cloves of garlic will you have left? Write and solve an equation.

   ____

Holt Mathematics

Problem Solving
3-12 *Solving Equations Containing Fractions*

Write the correct answer.

1. At the 2002 Winter Olympics, Austria won 2 gold medals. This was $\frac{1}{8}$ of the total medals Austria won. How many medals did Austria win?

2. At the 2002 Winter Olympics, Germany won 35 medals, of which 16 were silver. They won $1\frac{1}{3}$ times as many silver medals as gold medals. How many gold medals did Germany win?

3. Jesse plays ice hockey. He is on the ice $\frac{2}{5}$ of each game. A game lasts 45 minutes, divided into 3 periods. How many minutes is Jesse on the ice during each game?

4. Amelia's soccer team won $\frac{3}{4}$ of its games. The team won 18 games. How many games did the team play?

Choose the letter for the best answer.

5. At the 2002 Winter Olympics, China won 8 medals. The United States won $4\frac{1}{4}$ times as many medals as China did. How many medals did the United States win?

   A 24 medals   C 34 medals

   B 32 medals   D 40 medals

6. Juanita's water bottle contains $1\frac{7}{8}$ liters. She uses her bottle to fill Felix's. When she is done, her bottle contains $\frac{1}{4}$ liter. How many liters can Felix's bottle hold?

   F $\frac{7}{8}$ liters   H $1\frac{5}{8}$ liters

   G $1\frac{1}{8}$ liters   J $2\frac{1}{8}$ liters

7. Yoriko ran $6\frac{1}{2}$ laps of the track. Each lap is $\frac{1}{4}$ of a mile. How many miles did she run?

   A $6\frac{5}{8}$ miles   C 2 miles

   B $4\frac{1}{2}$ miles   D $1\frac{5}{8}$ miles

8. Rocky runs $3\frac{1}{2}$ miles each week. Leroy runs $5\frac{1}{3}$ miles each week. How much farther does Leroy run?

   F $1\frac{1}{2}$ miles   H $2\frac{1}{6}$ miles

   G $1\frac{5}{6}$ miles   J $2\frac{1}{4}$ miles

Holt Mathematics

LESSON 3-12 Reading Strategies
Compare and Contrast

Compare the steps for solving equations with fractions and solving equations with whole numbers.

Steps for Solving Equations		
	Whole Numbers	**Fractions**
Step 1: Get x by itself on one side of the equation.	$x - 8 = 7$	$x - \dfrac{3}{12} = \dfrac{4}{12}$
Step 2: Perform the opposite operation. In a subtraction problem, you add to get x by itself.	$x - 8 + 8 = 7 + 8$	$x - \dfrac{3}{12} + \dfrac{3}{12} = \dfrac{4}{12} + \dfrac{3}{12}$
Step 3: Solve.	$x = 15$	$x = \dfrac{7}{12}$

Use the chart to answer each question.

1. What is the first step to solve an equation with whole numbers?

____

2. Compare the first step in solving an equation with whole numbers to fractions. Is it the same or different?

____

3. What is the second step in solving an equation with whole numbers?

____

4. Compare the second step in solving an equation with whole numbers to an equation with fractions. Is it the same or different?

____

5. What is the opposite operation in both of the examples in the chart?

____

6. What is the third step in solving an equation with whole numbers?

____

7. Compare the third step in solving an equation with whole numbers to solving an equation with fractions. Is it the same or different?

____

Holt Mathematics

LESSON 3-12 Puzzles, Twisters & Teasers
What's Up, Doc?

Solve the equations. Then use the letters of the variables to answer the riddle.

1. $\frac{7}{18} + t = -\frac{4}{7}$ $t =$ _____________

2. $e - \frac{1}{5} = \frac{3}{5}$ $e =$ _____________

3. $g + \frac{11}{20} = \frac{3}{4}$ $g =$ _____________

4. $\frac{2}{3}k = \frac{4}{5}$ $k =$ _____________

5. $3a = \frac{6}{7}$ $a =$ _____________

6. $\frac{5}{12} + o = \frac{2}{3}$ $o =$ _____________

7. $m - \frac{1}{2} = \frac{1}{4}$ $m =$ _____________

8. $\frac{1}{5}r = 8$ $r =$ _____________

9. $s - \frac{2}{3} = \frac{5}{6}$ $s =$ _____________

10. $\frac{2}{3}x = \frac{3}{5}$ $x =$ _____________

Why did the rabbit buy a pure gold ring?

$\overline{\quad -\frac{121}{126} \quad}$ $\overline{\quad \frac{1}{4} \quad}$ $\overline{\quad \frac{1}{5} \quad}$ $\overline{\quad \frac{4}{5} \quad}$ $\overline{\quad -\frac{121}{126} \quad}$

$\overline{\quad \frac{3}{4} \quad}$ $\overline{\quad \frac{1}{4} \quad}$ $\overline{\quad 40 \quad}$ $\overline{\quad \frac{4}{5} \quad}$

$\overline{\quad 1\frac{1}{5} \quad}$ $\overline{\quad \frac{2}{7} \quad}$ $\overline{\quad 40 \quad}$ $\overline{\quad \frac{2}{7} \quad}$ $\overline{\quad -\frac{121}{126} \quad}$ $\overline{\quad 1\frac{1}{2} \quad}$

Holt Mathematics

Practice A
Estimate with Decimals

Round to the nearest whole number.

1. 4.23   2. 1.91   3. 10.75

    4   2   11

4. 5.88   5. 12.07   6. 18.70

    6   12   19

Estimate by rounding. Estimates may vary.

7. 8.4 + 15.9   8. 3.45 + 5.34   9. 9.36 + 7.542

    24   8   17

10. 9.6 + (−7.2)   11. −99.67 + −49.87   12. −4.5 + 6.2

    3   −150   1

13. 16.8 − 4.3   14. 34.25 − 18.5   15. 23.3 − 3.66

    13   15   19

Estimate using compatible numbers. Sample answers given.

16. 7.9 • 0.6   17. 6.35 • 1.83   18. 4.4 • 7.15

    8   12   28

19. 5.24 • 10.78   20. 29.1 • 5.15   21. 14.25 • 2.99

    55   150   45

22. 16.2 ÷ 3.5   23. 32.5 ÷ 3.2   24. 36.34 ÷ 2.1

    4   11   18

25. Cheri stopped for gasoline at a station that was charging $2.39 per gallon. If Cheri had $12.45 in cash, about how many gallons of gas could she buy?   **about 6 gallons**

3   **Holt Mathematics**

Practice B
Estimate with Decimals

Estimate by rounding to the nearest integer. Estimates may vary.

1. 7.45 + 35.84   2. 64.08 − 23.47   3. 6.842 + 14.05

    43   41   21

4. 7.156 + 8.34   5. 84.23 + (−78.24)   6. 3.78 − 2.078

    15   6   2

7. 46.47 − 98.75   8. 87.24 − 56.38   9. 6.324 + 60.324

    −53   31   66

10. −28.318 + 18.955   11. 35.082 + 8.37   12. −62.49 − 12.84

    −9   43   −75

Use compatible numbers to estimate. Sample answers given.

13. 59.69 ÷ 19.904   14. 86.234 • 9.876   15. 54.87 • 19.47

    3   860   1,100

16. −16.04 • 10.45   17. 31.25 • 6.57   18. 92.67 ÷ 32.89

    −160   210   3

19. 5.548 • 12.38   20. 88.42 ÷ 7.589   21. 90.05 ÷ 6.21

    72   11   15

22. Lisha works 20 hours per week at the bowling alley and makes $8.55 an hour. She gets a raise of $1.30 an hour. Approximately how much more will she make each week with her raise?

$20

23. Miguel is able to save $87.34 each month. He wants to buy a guitar that costs $542.45. For about how many months will Miguel have to save before he can buy the guitar?

about 6 months

4   **Holt Mathematics**

Practice C
Estimate with Decimals

Estimate. Estimates may vary. Sample answers given.

1. 6.187 + 61.871   2. 52.846 − (−5.284)   3. 77.435 ÷ 21.234

    68   58   4

4. 7.894 + (−7.894)   5. 84.324 • 9.846   6. 47.445 + 96.845

    0   840   144

7. −896.23 − 24.571   8. −7.501 • 67.499   9. 483.88 ÷ (−15.73)

    −921   −490   −30

10. 804.504 + 905.405   11. −46.781 − 5.23   12. 30.214 • 89.321

    1,710   −52   2,700

13. 98.346 + 162.54   14. 84.067 + (−101.26)   15. 720.089 ÷ 44.506

    261   −17   18

16. 645.645 • (−1.49)   17. 563.647 ÷ 27.804   18. −61.248 + (−65.234)

    −645   20   −126

19. 2.605 − 11.019   20. 40.93 • 15.042   21. 27.38 − (−6.27)

    −8   600   33

22. 125.88 ÷ 23.91   23. −38.5 ÷ 5.34   24. 59.64 • (−11.6)

    5   −8   −720

25. Niles is driving from New York City to Baltimore. He drives 64.8 mi/h for 2.9 hours. Approximately how far is Baltimore from New York City?

195 miles

26. Charlotte had $204. She bought a skirt for $78 and a scarf for $59. She wants to buy a pair of boots that costs $75. Does she have enough money left to buy the boots? About how much money does she have?

no; about $60

5   **Holt Mathematics**

Reteach
Estimate with Decimals

You can estimate with decimals by rounding each number to its greatest **place**.

Estimate: 52.38 + 9.006	**Estimate: 97.45 − 14.9**
The greatest place of 52.38 is **tens**.	The greatest place of 97.45 is **tens**.
The greatest place of 9.006 is **ones**.	The greatest place of 14.9 is **tens**.
52.38 → 50 9.006 → 9	97.45 → 100 14.9 → 10
50 + 9 = 59	100 − 10 = 90
So, 52.38 + 9.006 is about 59.	So, 97.45 − 14.9 is about 90.

Estimate: 16.35 • 1.8	**Estimate: 29.7 ÷ 3.65**
The greatest place of 16.35 is **tens**.	The greatest place of 29.7 is **tens**.
The greatest place of 1.8 is **ones**.	The greatest place of 3.65 is **ones**.
16.35 → 20 1.8 → 2	29.7 → 30 3.65 → 4
20 • 2 = 40	30 ÷ 4 is about 32 ÷ 4 = 8
So, 16.35 • 1.8 is about 40.	So, 29.7 ÷ 3.65 is about 8.

Estimate.

1. 7.843 + 54.1

   Greatest place of 7.843   **ones**

   7.843 rounds to   **8**

   Greatest place of 54.1   **tens**

   54.1 rounds to   **50**

   Estimate: **8 + 50 = 58**

2. 28.45 − 14.602

   Greatest place of 28.45   **tens**

   28.45 rounds to   **30**

   Greatest place of 14.602   **tens**

   14.602 rounds to   **10**

   Estimate: **30 − 10 = 20**

3. 6.41 • (−19.725)

   Greatest place of 6.41   **ones**

   6.41 rounds to   **6**

   Greatest place of −19.75   **tens**

   −19.725 rounds to   **−20**

   Estimate: **6(−20) = −120**

4. 44.67 ÷ 8.6

   Greatest place of 44.67   **tens**

   44.67 rounds to   **40**

   Greatest place of 8.6   **ones**

   8.6 rounds to   **9**

   40 ÷ 9 is about **45 ÷ 9**

   Estimate: **45 ÷ 9 = 5**

5. 36.5 + 78.09   6. 9.45 + (−2.75)   7. 98.56 − 53.381

    120   6   50

8. 33.52 • 5.29   9. −68.3 • 4.344   10. 24.65 ÷ 4.92

    150   −280   5

6   **Holt Mathematics**

Challenge
Just Estimate

Use estimates to compare: $8.62 - 15.91$? $-2.4(3.8)$

$$8.62 - 15.91 \quad ? \quad -2.4(3.8)$$

Think: $9 - 16 = -7$ Think: $-2 \cdot 4 = -8$

$-7 > -8$, so $8.62 - 15.91 > -2.4(3.8)$

Estimate to compare. Use > or <.

1. $3.45 + 11.65$ $>$ $16.8 - 4.1$

2. $-0.54(14.22)$ $>$ $-23.65 + 9.44$

3. $-2.59 - 5.18$ $<$ $6.12 - 13.3$

4. $-9.6(-7.25)$ $<$ $82.45 - 10.3$

5. $24.7 \div 4.5$ $<$ $12.8 - 5.3$

6. $-20.8 + 12.3$ $>$ $-20.2 \div 1.8$

7. $125.95 - 25.4$ $>$ $1.5(50.38)$

8. $-14.57 - 7.34$ $<$ $-39.62 \div 2.3$

9. $4.9(-11.6)$ $<$ $40.29 + 20.51$

10. $47.6 - 17.7$ $<$ $-7.7(-4.1)$

11. $23.6 \div 5.5$ $<$ $13.08 - 8.4$

12. $9.83 - 23.41$ $<$ $-6.42 - 5.19$

13. $18.45 \div 1.82$ $<$ $49.5 \div 5.46$

14. $0.35(-3.55)$ $<$ $4.68 - 5.54$

15. $32.7 \div (-2.6)$ $<$ $-5.28(1.52)$

16. $-5.49(-4.06)$ $>$ $24.62 - 6.4$

17. $-3.24 - 16.4$ $>$ $-6.6(2.8)$

18. $0.78(56.5)$ $<$ $-9.1 + 66.65$

19. $-45.3 \div 4.55$ $>$ $79.9 \div (-8.48)$

20. $-6.7(-5.08)$ $<$ $18.5(1.54)$

21. $65.36 \div (-10.7)$ $>$ $-9.3 + 1.48$

22. $-3.53(-2.45)$ $<$ $-119.5 \div (-11.7)$

23. $65.5 - 67.09$ $<$ $34.4 \div 33.55$

24. $-40.06(-3.1)$ $>$ $97.8 + 19.7$

7

Problem Solving
Estimate with Decimals

Write the correct answer. Estimates may vary. Sample answers given.

1. The Spanish Club makes a profit of $1.85 on every pie sold at a bake sale. The goal is to earn $40.00 selling pies. Will the club have to sell more than or fewer than 20 pies to meet the goal?

more than 20 pies

2. Murray and 4 friends split the cost of a pizza, with each paying the same amount. Murray has $3.36. The pizza costs $14.20. Does Murray have enough to pay for his share? About how much is his share?

yes; about $3.00

3. Luis has 22 MB of free space on his MP3 player. He wants to download 5 songs. They will take up 5.1, 4.1, 4.3, 8.2, and 3.6 MB of space. Does Luis have enough free space? About how much space is needed?

no; about 25 MB

4. Debi has a gift certificate worth $50 for a local book-and-music store. She decides to buy a book that costs $18.95 and some CDs. Each CD costs $13.99. How many CDs can she buy?

2 CDs

Choose the letter for the best answer.

This table shows the number of dollars foreign tourists spent while visiting different countries in 2003.

Spending by Tourists in 2003

Country	Amount Spent by Foreign Tourists ($ billions)
United States	65.1
Spain	41.7
France	36.3
Italy	31.3
Germany	22.8
United Kingdom	19.5
China	17.4
Austria	13.6
Turkey	13.2
Greece	10.6
Mexico	9.5

5. The amount spent in which two countries was about equal to the amount spent in Italy?
   A Austria and Turkey
   B United Kingdom and China
   C Germany and Greece
   D Austria and China

6. About how much did tourists spend in Spain, France, and Italy all together?
   F $111 billion   H $78 billion
   G $109 billion   J $67 billion

7. If the amount spent by foreign tourists remains the same, about how much will foreign tourists spend in the United States over 5 years?
   F $120 billion   H $350 billion
   G $250 billion   J $450 billion

8

Reading Strategies
Make Generalizations

Estimating is useful when you don't need an exact answer. Knowing how to round decimals helps you estimate.

Rules for Rounding Decimals
1. Look at the digit in the tenths place.
2. If that digit is five or greater, round the number in the ones place up one.
3. If that digit is less than five, round the number in the ones place down one.

Estimate the sum. Use the rules in the table to answer each question.

$$\begin{array}{r} 7.48 \\ + 2.68 \\ \hline \end{array}$$

1. In the number 7.48, will you round 7 up or down? Explain.

round down, because 4 is less than 5

2. In the number 2.68, will you round 2 up or down? Explain.

round up, because 6 is greater than 5

3. Rewrite the equation and solve using the rounded numbers.

$7 + 3 = 10$

Use the rounding rules to estimate the difference.

$$\begin{array}{r} 13.78 \\ - 5.23 \\ \hline \end{array}$$

4. What is 13.78 rounded to the nearest whole number?

14

5. What is 5.23 rounded to the nearest whole number?

5

6. Is it easier to solve $13.78 - 5.23$ or $14 - 5$? Explain.

It is easier to solve $14 - 5$ because there are fewer digits to subtract.

9

Puzzles, Twisters & Teasers
Round and Round We Go!

Estimate sums and differences by rounding. Estimate products and quotients by using compatible numbers. Then solve the riddle.

I $-9.916 + 12.4$ __2__

D $24.79 \cdot 9.83$ __250__

L $37.2 + 25.83$ __63__

N $-36.8 + 14.217$ __−23__

G $68.2 + 23.67$ __92__

U $61.45 \div 9.08$ __7__

O $15 - 6.835$ __8__

R $37.63 \div 7.43$ __5__

A $5.921 - 13.2$ __−7__

H $98.6 + 43.921$ __143__

E $62.84 - 35.169$ __28__

Y $-3.96 \cdot 14.81$ __−60__

What did the hat say to the tie?

I	L	L	G	O	O	N
2	63	63	92	8	8	−23

A	H	E	A	D,	Y	O	U
−7	143	28	−7	250	−60	8	7

H	A	N	G
143	−7	−23	92

A	R	O	U	N	D.
−7	5	8	7	−23	250

10

Practice A
Adding and Subtracting Decimals

Add. Estimate to check whether your answer is reasonable.

1. 3.52 + 6.33

Tens	Ones	.	Tenths	Hundredths
	3	.	5	2
+	6	.	3	3
	9	.	8	5

2. 9.48 + 12.75

Tens	Ones	.	Tenths	Hundredths
	9	.	4	8
+	1	2	.	7	5
2	2	.	2	3

3. 7.97 + 3.6

Tens	Ones	.	Tenths	Hundredths
	7	.	9	7
+	3	.	6	
1	1	.	5	7

4. 13 + (−10.35)

Tens	Ones	.	Tenths	Hundredths
1	3	.		
+	−1	0	.	3	5
	2	.	6	5

5. 0.82 + 1.54

2.36

6. 12.28 + 19.019

31.299

7. 29.5 + 3.676

33.176

Subtract. Estimate to check whether your answer is reasonable.

8. 24.6 − 17.93

Tens	Ones	.	Tenths	Hundredths
2	4	.	6	
−	1	7	.	9	3
	6	.	6	7

9. 40 − 19.4

Tens	Ones	.	Tenths	Hundredths
4	0	.		
−	1	9	.	4	
2	0	.	6	

10. 8.4 − 4.6

3.8

11. 9.2 − 3.8

5.4

12. 13.4 − 8.2

5.2

13. 15.42 − 8.3

7.12

14. 9 − 5.5

3.5

15. 21.68 − 12

9.68

11
Holt Mathematics

Practice B
Adding and Subtracting Decimals

Add. Estimate to check whether each answer is reasonable.

1. 6.14 + 8.91

15.05

2. 4.51 + 13.08

17.59

3. 12.54 + 21.08

33.62

4. 34.22 + (−18.5)

15.72

5. −10.10 + (−5.9)

−16

6. 6.87 + (−31.6)

−24.73

7. 9 + 5.68

14.68

8. −15.51 + 8.55

−6.96

9. 36.36 + 54.54

90.9

Subtract.

10. 6.23 − 3.62

2.61

11. 8.67 − 6.87

1.8

12. 28.94 − 9.48

19.46

13. 23.57 − 6.84

16.73

14. 16.61 − 7.56

9.05

15. 32.08 − 12.37

19.71

16. 19 − 6.92

12.08

17. 42 − 31.89

10.11

18. 23 − 21.45

1.55

19. 46.2 − 0.27

45.93

20. 22 − 18.63

3.37

21. 58.9 − 29.58

29.32

22. Anna swims the length of the pool in 38.45 seconds and then swims the length of the pool again in 42.38 seconds. What is her total time for 2 lengths of the pool?

80.83 seconds

23. Po has 2 gerbils named Yip and Yap. Yip weighs 3.62 ounces, and Yap weighs 2.79 ounces. How much heavier is Yip than Yap?

0.83 ounce

12
Holt Mathematics

Practice C
Adding and Subtracting Decimals

Add or subtract. Estimate to check whether each answer is reasonable.

1. 8.326 + 8.49

16.816

2. 13.845 − 10.87

2.975

3. 42.35 + 4.368

46.718

4. 8.695 − 7.52

1.175

5. −5.421 + 18.4

12.979

6. −12.54 − 16.951

−29.491

7. 6.742 − 9.458

−2.716

8. 8.552 + 14.84

23.392

9. −75.25 + (−6.382)

−81.632

10. 47.68 − 9.654

38.026

11. −38.59 − (−7.816)

−30.774

12. 9.561 + (−16.82)

−7.259

13. 18.234 + 6.357

24.591

14. 123.45 − 12.345

111.105

15. 6.456 + (−87.51)

−81.054

16. 74.832 − 56.842

17.99

17. 95.216 − 59.162

36.054

18. 82.64 + (−45.846)

36.794

19. 27.205 + 19.17 + 8.357

54.732

20. 12.63 + 72.029 + 5.9

90.559

21. 49.408 + 56.27 − 13.02 − 25.821

66.837

22. 9.30 + 17.43 + 38.57 − 25.719

39.581

23. A whippet can run a 200-yard course at a rate of 35.5 mi/h. This is 3.85 mi/h slower than the top speed of a greyhound. How fast can a greyhound run?

39.35 mi/h

24. Heike has 2 rocks. If the total weight of both rocks is 54.283 kilograms and the first rock weighs 29.928 kilograms, how much does the second rock weigh?

24.355 kilograms

13
Holt Mathematics

Reteach
Adding and Subtracting Decimals

You can use a place-value chart to add decimals.

Add: 3.48 + 2.7

Step 1: Line up the decimal points. Use 0 as a placeholder for hundredths in 2.7.

Ones	.	Tenths	Hundredths
	.		
3	.	4	8
+ 2	.	7	0

Subtract: 68 − 5.9

Step 1: Line up the decimal points. Use 0 as a placeholder for tenths in 68.

Tens	Ones	.	Tenths
		.	
6	8	.	0
−	5	.	9

Step 2: Add. Place the decimal point in the answer.

Ones	.	Tenths	Hundredths
1			
3	.	4	8
+ 2	.	7	0
6	.	1	8

Step 2: Subtract. Place the decimal point in the answer.

Tens	Ones	.	Tenths
	7		10
6	8̸	.	0̸
−	5	.	9
6	2	.	1

Step 3: Estimate: 3 + 3 = 6. So, 6.18 is a reasonable answer.

Step 3: Estimate to check: 68 − 6 = 62. So, 62.1 is a reasonable answer.

Add or subtract. Estimate to check your answer.

1. 19.67 + 8.45

Tens	Ones	.	Tenths	Hundredths
1	9	.	6	7
+	8	.	4	5
2	8	.	1	2

2. 9.36 − 7.29

Tens	Ones	.	Tenths	Hundredths
	9	.	3	6
−	7	.	2	9
	2	.	0	7

3. 25.36 − 9.7

Tens	Ones	.	Tenths	Hundredths
2	5	.	3	6
−	9	.	7	0
1	5	.	6	6

4. 12.89 + 37.07

Tens	Ones	.	Tenths	Hundredths
1	2	.	8	9
+ 3	7	.	0	7
4	9	.	9	6

5. 34.71 + 9.19

43.9

6. 57.06 − 27.38

29.68

7. 48.3 + (−10.73)

37.57

14
Holt Mathematics

Challenge
Answer Match

Add or subtract. Draw lines to match equivalent answers in each column. Each answer in the middle column matches an answer in both outside columns.

1. $-61.244 + 59.014$	2. $21.65 + 14.2 + 10.03$	3. $41.729 + 14.92$
-2.23	45.88	59.649

4. $4.018 + 7.052$	5. $212.8 - 136.2$	6. $42.25 - 13.174$
11.07	76.6	29.076

7. $-104.02 + 149.9$	8. $0.015 - 6.72$	9. $-0.673 + (-5.707)$
45.88	-6.705	-6.38

10. $-5.23 - 0.804 + 35.11$	11. $-70.6 + 48.05 + 33.62$	12. $22.015 + 23.865$
29.076	11.07	45.88

13. $60.25 - 0.601$	14. $65.109 - 5.46$	15. $7.17 - 9.4$
59.649	59.649	-2.23

16. $-1.08 + 3.797 - 9.422$	17. $10.25 - 12.03 - 0.45$	18. $16.02 + 44.48 + 16.1$
-6.705	-2.23	76.6

19. $15.25 + 4.03 + 17.706$	20. $130.82 - 101.744$	21. $-5.61 - 1.095$
36.986	29.076	-6.705

22. $11.691 - 21.1 + 3.029$	23. $45.006 + (-8.02)$	24. $52.73 - 53.71 + 12.05$
-6.38	36.986	11.07

25. $-33.42 - (-110.02)$	26. $46.01 + 9.9 - 62.29$	27. $6.007 + 74.2 - 43.221$
76.6	-6.38	36.986

15

Problem Solving
Adding and Subtracting Decimals

Write the correct answer.

1. In the Pacific Ocean, the Philippine Trench is 10.05 kilometers deep. In the Atlantic Ocean, the Brazil Basin is 6.12 kilometers deep. How much deeper is the Philippine Trench than the Brazil Basin?

__3.93 kilometers__

2. Hawaii's Mauna Kea measures 9.75 kilometers from its base to its peak. The base of Mauna Kea lies 5.55 kilometers below the ocean. What is the height of the part of Mauna Kea that is above sea level?

__4.2 kilometers__

3. A team of mountain climbers makes camp 1.48 kilometers above sea level. They climb another 2.91 kilometers to the mountain's peak. How tall is the mountain?

__4.39 kilometers__

4. At dawn, the temperature at the summit of a mountain was $-8.5°C$. By noon, the temperature had increased $3.6°C$. What was the temperature at noon?

__$-4.9°C$__

Choose the letter for the best answer.

This table gives the heights of the tallest mountains on each of the seven continents.

The Seven Summits

Mountain	Country	Height (km)
Mt. Everest	Nepal-Tibet	8.85
Mt. Aconcagua	Argentina	6.96
Mt. McKinley	United States	6.19
Mt. Kilimanjaro	Tanzania	5.90
Mt. Elbrus	Russia	5.64
Vinson Massif	Antarctica	4.90
Puncak Jaya	New Guinea	4.88

5. In 2000, Joby Ogwyn became the youngest person to climb each of the seven summits. How much higher did he climb on Vinson Massif than on Puncak Jaya?
   - (A) 0.02 km
   - B 0.12 km
   - C 0.18 km
   - D 9.78 km

6. Mt. Kilimanjaro, Mt. Elbrus, and Mt. Aconcagua were the first three of the seven summits Joby climbed. What was the total height he climbed?
   - B 16.2 km
   - G 17.5 km
   - (H) 18.5 km
   - J 22 km

7. Mt. Aconcagua is about 1.3 kilometers taller than which mountain?
   - A Mt. Everest
   - (B) Mt. Elbrus
   - C Mt. McKinley
   - D Vinson Massif

16

Reading Strategies
Use a Grid

A grid is useful for adding and subtracting decimals.

Add 19.2 + 7.54.

Step 1: Make a grid that has enough squares for each number and one for the decimal point.

Step 2: Place one number in each square of the grid. Carefully line up the decimals.

1	9	.	2	
+	7	.	5	4

Step 3: If there is a square without a number, add a zero as a placeholder.

1	9	.	2	0
+	7	.	5	4

Use the grids to answer each question.

1. How do you place the numbers in a grid?

__One number is placed in each square.__

2. Why was a zero added to 19.2?

__as a placeholder to show that there is no number in that place__

3. Write the problem $40.3 - 6.54$ in the grid.

4	0	.	3	0
−	6	.	5	4

4. Did you need to add a zero as a placeholder? If so, where?

__yes; in the hundredths place of the first number__

5. Subtract 6.54 from 40.30 using a grid. What is the answer?

__33.76__

17

Puzzles, Twisters & Teasers
Up Against the Wall!

Add or subtract to find the solution. Round your solution to the nearest integer. Then answer the riddle.

R $5.37 + 16.45$ __22__

I $8.89 - 5.91$ __3__

E $7 + 5.82$ __13__

L $18.31 - 8.66$ __10__

N $4.97 - 3.2$ __2__

M $7.82 + 31.23$ __39__

O $6 + 9.33$ __15__

Y $5.98 + 12.99$ __19__

C $5.02 + 3.21$ __8__

U $5 - 0.53$ __4__

H $-7.23 + 6.9$ __0__

A $4.16 - 9.04$ __-5__

T $-32.7 + 62.82$ __30__

Whet did one wall say to the other wall?

I	L	L		M	E	E	T
3	10	10		39	13	13	30

Y	O	U		A	T		T	H	E
19	15	4		−5	30		30	0	13

C	O	R	N	E	R
8	15	22	2	13	22

18

Practice A
Multiplying Decimals

Multiply. Choose the letter for the best answer.

1. 5 • 0.05
- A 25
- B 2.5
- C 0.25
- D 0.025

2. 9 • 0.7
- F 63
- G 6.3
- H 0.63
- J 0.063

3. 6 • 0.003
- A 18
- B 1.8
- C 0.18
- D 0.018

4. 5 • 1.2
- F 60
- G 6
- H 0.6
- J 0.06

5. 6 • 1.8
- A 10.8
- B 1.08
- C 0.108
- D 0.0108

6. 8 • 0.02
- F 16
- G 1.6
- H 0.16
- J 0.016

7. 3 • 8.4
- A 25.2
- B 2.52
- C 0.252
- D 0.0252

8. 7 • 0.51
- F 357
- G 35.7
- H 3.57
- J 0.357

Multiply. Estimate to check whether each answer is reasonable.

9. 6.8 • 4   27.2

10. 8.1 • (−2)   −16.2

11. 9.5 • 5   47.5

12. 3.5 • 7   24.5

13. −6.3 • 6   −37.8

14. 9 • 3.7   33.3

15. −6.7 • (−5)   33.5

16. 8.8 • (−8)   −70.4

17. 5.2 • (−4)   −20.8

18. −3 • 4.1   −12.3

19. 1.5 • 1.2   1.8

20. −2.3 • 1.7   −3.91

21. Cecile walked 3.7 miles each day for 8 days last month. How many miles total did Cecile walk last month?

29.6 miles

Holt Mathematics

Practice B
Multiplying Decimals

Multiply.

1. 6 • 0.3   1.8

2. 3 • 0.05   0.15

3. 0.7 • 4   2.8

4. 8 • 6.1   48.8

5. 7.4 • 6   44.4

6. 1.4 • 9   12.6

7. 4.8 • 7   33.6

8. 3 • 8.2   24.6

9. 5.5 • 8   44

10. 1.5 • 6   9

11. 7.9 • 2   15.8

12. 5 • 6.9   34.5

Multiply. Estimate to check whether each answer is reasonable.

13. 6.3 • 7.8   49.14

14. 9.7 • (−4.7)   −45.59

15. 6.8 • 0.9   6.12

16. 2.8 • 8.2   22.96

17. −7 • 6.42   −44.94

18. 1.9 • 7.22   13.718

19. −5.3 • (−8.4)   44.52

20. 7.16 • 0.03   0.2148

21. 1.56 • (−7.8)   −12.168

22. 4.6 • 3.1   14.26

23. 0.62 • 1.45   0.899

24. −5.74 • 1.9   −10.906

25. Jordan jogged 4.8 miles each day for 21 days last month. How many miles did she jog last month?

100.8 miles

Holt Mathematics

Practice C
Multiplying Decimals

Multiply. Estimate to check whether each answer is reasonable.

1. 6.23 • 5.48   34.1404

2. 7.82 • 4.18   32.6876

3. 9.65 • 4.15   40.0475

4. 2.57 • 3.46   8.8922

5. 7.84 • 9.23   72.3632

6. 4.28 • 6.94   29.7032

7. 42.16 • 8.52   359.2032

8. 35.84 • 2.57   92.1088

9. 10.68 • 71.82   767.0376

10. 6.52 • 3.15   20.538

11. 6.78 • (−4.51)   −30.5778

12. 7.75 • 7.75   60.0625

13. 9.49 • 5.25   49.8225

14. −6.84 • 9.44   −64.5696

15. 10.58 • (−2.54)   −26.8732

16. −67.25 • (−5.61)   377.2725

17. 94.21 • (−7.81)   −735.7801

18. 5.27 • (−18.94)   −99.8138

19. −1.05 • 3.8 • 7.1   −28.329

20. 0.45 • 1.8 • 0.06   0.0486

21. 5.7 • 4.52 • (−8.3)   −213.8412

22. 0.05 • 1.7 • (−12.62)   −1.0727

23. 4.02 • (−9.9) • 23.7   −943.2126

24. −2.2 • 18.45 • (−0.8)   32.472

25. Phone service costs $24.00 per month plus $0.09 per minute for long distance calls. What is your monthly bill if you make 57 minutes of long distance calls? $29.13

26. A soccer uniform costs $3.98 for the shirt and $4.49 for the shorts. What is the total cost of 5 uniforms? $42.35

Holt Mathematics

Reteach
Multiplying Decimals

To multiply two decimals:

Step 1: Round each number to the nearest integer.
Step 2: Multiply the integers to estimate the product.
Step 3: Multiply the decimals.
Step 4: Place the decimal point in the product to make it closest to the estimate.

Multiply: 2.7 • 4.3

```
   4.3
   2.7
  301
  860
 11.61
```

11.61 is close to 12.

Think:
2.7 rounds to 3.
4.3 rounds to 4.
3 • 4 = 12
Place the decimal point in the product to make it closest to 12.

Multiply.

1. 6.7 • 9.1
- 6.7 rounds to 7
- 9.1 rounds to 9
- The product is close to 63.
- Product: 60.97

2. −3.21 • 8.8
- −3.21 rounds to −3
- 8.8 rounds to 9
- The product is close to −27.
- Product: −28.248

3. 4.1 • 0.8
- 4.1 rounds to 4
- 0.8 rounds to 1
- The product is close to 4.
- Product: 3.28

4. 12.3 • (−2.7)
- 12.3 rounds to 12
- −2.7 rounds to −3
- The product is close to −36.
- Product: −33.21

Multiply. Estimate to place the decimal point.

5. 2.06 • 7.9   16.274

6. −4.89 • 0.6   −2.934

7. 8.23 • (−4.2)   −34.566

Holt Mathematics

Holt Mathematics

Challenge
Decimal Round

Think about numbers that could have resulted in this estimate.

$$0.8 \cdot 3 = 2.4$$

The *least possible factors* that could have been rounded up are 0.75 and 2.5. So, the *least product* is $0.75 \cdot 2.5 = 1.875$.

The *greatest possible factors* that could have been rounded down are 0.84 and 3.4. So, the *greatest product* is $0.84 \cdot 3.4 = 2.856$.

Find the products of the rounded numbers below. Write the least and greatest factors for each product before rounding. Then find the least and greatest products.

	Rounded Product	Least Product	Greatest Product
1.	$18 \cdot 12 = 216$	$17.5 \cdot 11.5 = 201.25$	$18.4 \cdot 12.4 = 228.16$
2.	$3.3 \cdot 4 = 13.2$	$3.25 \cdot 3.5 = 11.375$	$3.34 \cdot 4.4 = 14.696$
3.	$0.9 \cdot 0.7 = 0.63$	$0.85 \cdot 0.65 = 0.5525$	$0.94 \cdot 0.74 = 0.6956$
4.	$1.8 \cdot 2.2 = 3.96$	$1.75 \cdot 2.15 = 3.7625$	$1.84 \cdot 2.24 = 4.1216$
5.	$6 \cdot 0.1 = 0.6$	$5.5 \cdot 0.05 = 0.275$	$6.4 \cdot 0.14 = 0.896$
6.	$4.5 \cdot 0.6 = 2.7$	$4.45 \cdot 0.55 = 2.4475$	$4.54 \cdot 0.64 = 2.9056$
7.	$10 \cdot 1.1 = 11$	$9.5 \cdot 1.05 = 9.975$	$10.4 \cdot 1.14 = 11.856$
8.	$0.3 \cdot 15 = 4.5$	$0.25 \cdot 14.5 = 3.625$	$0.34 \cdot 15.4 = 5.236$
9.	$8.7 \cdot 5.1 = 44.37$	$8.65 \cdot 5.05 = 43.6825$	$8.74 \cdot 5.14 = 44.9236$
10.	$1.4 \cdot 1.9 = 2.66$	$1.35 \cdot 1.85 = 2.4975$	$1.44 \cdot 1.94 = 2.7936$

23
Holt Mathematics

Problem Solving
Multiplying Decimals

Write the correct answer.

1. A group of 6 adults bought tickets to a play at the community center. The tickets cost $17.50 each. How much did the tickets cost in all?

   $105

2. Student tickets for the basketball playoffs cost $9.25 each. How much do 9 student tickets cost?

   $83.25

3. Movie tickets for senior citizens cost 0.8 times as much as regular adult tickets. Adult tickets cost $7.50. How much do senior citizen tickets cost?

   $6.00

4. About 30.5 million Americans attend classical music concerts. The average concertgoer attends 2.9 concerts per year. About how many tickets to classical concerts are sold each year?

   about 88.45 million tickets

Choose the letter for the best answer.

5. The population of Illinois in 2000 was about 1.57 times its population in 1940. If the population of Illinois in 1940 was about 7.9 million, what was its population in 2000, to the nearest tenth of a million?

   A 1.2 million
   B 9.5 million
   C 12.4 million
   D 15.8 million

6. In 2004, the United States had 32.5 million broadband subscribers. This number is expected to increase 1.75 times by 2008. How many broadband subscribers is the United States expected to have in 2008?

   F 24.375 million
   G 34.25 million
   H 56.875 million
   G 568.75 million

7. Nauru and Gibraltar are among the smallest countries in the world. Nauru has about 3.28 times the area that Gibraltar does. Gibraltar is 2.5 square miles. What is the area of Nauru?

   A 0.78 mi^2
   B 0.82 mi^2
   C 5.78 mi^2
   D 8.2 mi^2

8. From 1999 to 2008, the United States Mint is issuing a series of 50 quarters representing each of the 50 states. What is the cost of collecting 10 of each quarter?

   F $100
   G $125
   H $150
   J $250

24
Holt Mathematics

Reading Strategies
Compare and Contrast

Decimals are multiplied in much the same way that you multiply whole numbers.

Multiply Whole Numbers	Multiply Decimals
5	0.5
× 7	× 0.7
35	0.35

Compare multiplying whole numbers to multiplying decimals.

1. What is the same about multiplying whole numbers and decimals?

   The same numerals are in both answers.

2. What is different about multiplying whole numbers and decimals?

   There are decimal places in the product when you multiply decimals, but none with whole numbers.

It is important to place the decimal point correctly in the product.

Steps for Placing the Decimal Point in the Product	Example: 1.37×0.8
Step 1: Find the product.	1096
Step 2: Count the number of decimal places in each factor.	1.37 → 2 places 0.8 → 1 place
Step 3: Find the total number of decimal places in both numbers.	3 places
Step 4: Using the number found in Step 3, move that number of places to the left in the product and place the decimal point.	1.096

3. How many decimal places are in 0.63?

   2

4. How many decimal places are in 4.231?

   3

5. How many decimal places will be in the product of 0.63×4.231?

   5

25
Holt Mathematics

Puzzles, Twisters & Teasers
Turning of the Times!

Multiply the decimals to solve. Round your solution to the nearest integer. Then answer the riddle.

S $9 \cdot 0.36$ — 3
O $8 \cdot 0.6$ — 5
U $6 \cdot 4.9$ — 29
T $2.4 \cdot 3.2$ — 8
I $3.0 \cdot 3.2$ — 10
R $0.6 \cdot 0.05$ — 0
N $3 \cdot 7.01$ — 21
A $-2.6 \cdot 0.4$ — -1
G $8 \cdot 0.77$ — 6
H $1.7 \cdot 12$ — 20
E $2.8 \cdot -1.6$ — -4
W $7 \cdot 0.3$ — 2

When is a car not a car?

W	H	E	N		I	T
2	20	−4	21		10	8

T	U	R	N	S
8	29	0	21	3

I	N	T	O		A
10	21	8	5		−1

G	A	R	A	G	E
6	−1	0	−1	6	−4

26
Holt Mathematics

Holt Mathematics

Practice A
Dividing Decimals by Integers

Match each division with the correct quotient.

1. $4.5 \div 3$ **A.** 0.8
2. $6.4 \div 8$ **B.** 1.21
3. $8.1 \div 9$ **C.** 1.15
4. $4.84 \div 4$ **D.** 1.5
5. $1.68 \div 8$ **E.** 0.9
6. $5.04 \div 7$ **F.** 0.91
7. $6.9 \div 6$ **G.** 0.21
8. $4.55 \div 5$ **H.** 0.72

9. $2.79 \div 3$ **R.** 2.4
10. $9.45 \div (-7)$ **S.** 0.93
11. $21.6 \div 9$ **T.** 12.3
12. $-2.5 \div 5$ **U.** −1.35
13. $24.6 \div 2$ **V.** 0.59
14. $7.7 \div (-11)$ **W.** −1.31
15. $5.9 \div 10$ **X.** −0.5
16. $-10.48 \div 8$ **Y.** −0.7

Divide. Estimate to check whether each answer is reasonable.

17. $5.4 \div 9$ 0.6
18. $0.66 \div 3$ 0.22
19. $15.3 \div (-6)$ −2.55

20. $-2.8 \div 14$ −0.2
21. $13.04 \div 8$ 1.63
22. $-5.5 \div 11$ −0.5

23. Alyssa and Kim bought a new volleyball for $12.95 and a new net for $16.75. They shared the cost equally. How much did they each pay? $14.85

24. Dominic's scores for his skating performance were 5.5, 5.2, and 5.5. What was his average score?   5.4

27   **Holt Mathematics**

Practice B
Dividing Decimals by Integers

Divide. Estimate to check whether each answer is reasonable.

1. $20.8 \div 8$ 2.6
2. $54.4 \div 5$ 10.88
3. $0.876 \div 6$ 0.146

4. $65.6 \div 4$ 16.4
5. $-96.88 \div 7$ −13.84
6. $50.4 \div 18$ 2.8

7. $67.42 \div 4$ 16.855
8. $88.65 \div (-3)$ −29.55
9. $77.25 \div 5$ 15.45

10. $-0.18 \div 4$ −0.045
11. $41.17 \div (-23)$ −1.79
12. $74.55 \div 25$ 2.982

13. $0.144 \div 4$ 0.036
14. $5.36 \div (-8)$ −0.67
15. $27.6 \div 12$ 2.3

16. $22.08 \div (-3)$ −7.36
17. $1.976 \div 13$ 0.152
18. $25.56 \div (-5)$ −5.112

19. $0.504 \div 9$ 0.056
20. $170.1 \div 27$ 6.3
21. $5.25 \div (-3)$ −1.75

22. Doris collects wicker baskets. She spent $9.56 on 3 baskets at the flea market. Then she found 4 more baskets at a garage sale. She paid $10.67 for those baskets. What was the average price per basket for all 7 baskets? $2.89

23. As of January 2002, 3 top college football coaches had the following winning percents in bowl games: 0.740, 0.692, and 0.683. What was their average winning percent?   0.705

28   **Holt Mathematics**

Practice C
Dividing Decimals by Integers

Divide. Estimate to check whether each answer is reasonable.

1. $86.24 \div 8$ 10.78
2. $56.64 \div 4$ 14.16
3. $0.735 \div 7$ 0.105

4. $75.75 \div 5$ 15.15
5. $92.7 \div 9$ 10.3
6. $243.25 \div 7$ 34.75

7. $83.61 \div 3$ 27.87
8. $312.85 \div 5$ 62.57
9. $-0.846 \div 18$ −0.047

10. $67.656 \div 12$ 5.638
11. $395.55 \div (-15)$ −26.37
12. $10.353 \div 21$ 0.493

13. $581.98 \div (-14)$ −41.57
14. $-291.89 \div 17$ −17.17
15. $122.21 \div 11$ 11.11

Simplify each expression.

16. $2.8 - 14.07 \div 3 + 6.2$ 4.31
17. $26.65 \div 4.1 + 8.047$ 14.547

18. $20 \cdot 32.4 \div 6 \times 5$ 540
19. $1.22 \cdot 1.1 \div (-2) - 0.57$ −1.241

20. $(91.89 + 10.51) \div 16 + 1.045$ 7.445
21. $67.44 \div 24 + 0.23 - 14.076$ −11.036

22. An investor pays $478.75 to purchase 25 shares of stock. The total price includes a $35 commission. What is the cost of each share of stock? $17.75

23. Quinn swam three laps during her swimming class. If her times were 32.02 seconds, 33.48 seconds, and 32.96 seconds, what was her average lap time?   32.82 s

29   **Holt Mathematics**

Reteach
Dividing Decimals by Integers

You can use models to divide decimals by integers.

Divide $0.35 \div 5$.

Represent 0.35 by shading 35 out of 100 squares.

Divide the shaded area into 5 equal parts.

Each part is made up of 7 squares. This represents 0.07.

$0.35 \div 5 = 0.07$ or $5\overline{)0.35}$

Notice that the decimal point in the quotient is directly above the decimal point in the dividend.

Use the model to divide.

1. $0.48 \div 4$ 0.12
2. $1.8 \div 3$ 0.6
3. $0.16 \div 4$ 0.04

Divide. Draw models to help you.

4. $0.56 \div 8$ 0.07
5. $0.75 \div 5$ 0.15
6. $7.2 \div 9$ 0.8

7. $10.5 \div 5$ 2.1
8. $6.4 \div 16$ 0.4
9. $7.08 \div 6$ 1.18

30   **Holt Mathematics**

Challenge
Code to Order

Simplify each expression. Write the answers in order from least to greatest on the top lines of the certificate below. Then write the letter that corresponds to each answer in the same order on the bottom lines to see a secret message.

1. $5.43 + 8(6 - 2.7) - 0.3^2 \div 9 + 12.6$ ___**44.42**___ [O]

2. $4^2(2 - 0.5) - 6.08 \div 2^3 + 13 \cdot 4.2$ ___**77.84**___ [O]

3. $(7.4 + 2.1) \div 5 + 1.2 \cdot (9 \div 0.3) - 8.4$ ___**29.5**___ [E]

4. $126 \div 12 \cdot 2 - 8(0.2 + 5) \div 2$ ___**0.2**___ [A]

5. $4 \div 0.2^3 \div 5 + (-5.95 + 5.45)^2$ ___**100.25**___ [B]

6. $72.32 \div 4 - 0.2^2 + 2.1 \div 0.3$ ___**25.04**___ [W]

7. $(0.2 \cdot 40 \div 2^2 + 10) \div 0.2^2 \cdot 0.5^2$ ___**75**___ [J]

8. $8(32 \div 1.6) \div 0.8 \cdot (0.7^2 + 0.01) \div 2.5$ ___**40**___ [S]

9. $54.18 + 2.34 \div 3.9 - 2.1 \cdot 1.2$ ___**52.26**___ [M]

10. $8.35 \div 5 + 5(4.86 \div 1.8)^2 \cdot 2$ ___**74.57**___ [E]

0.2	25.04	29.5	40	44.42	52.26	74.57	75	77.84	100.25
A	W	E	S	O	M	E	J	O	B

Holt Mathematics

Problem Solving
Dividing Decimals by Integers

Write the correct answer.

1. Teri collects loose change in 3 cans placed near cash registers at the mall. One can holds $37.18. The second can holds $44.25. The third can holds $50.84. If Teri divides the money equally among 3 charities, how much money will each charity get?

   $44.09

2. In 2003, listeners in the United States paid $0.07 billion to download music. The expected increase in sales of downloaded music is $2.19 billion by 2008. What is the average increase in sales expected over each of those 5 years?

   $0.424 billion

3. The longest life expectancies are found in Okinawa (81.2 years), Japan (79.9 years), and Hong Kong (79.1 years). Life expectancy in the United States is 76.8 years. About how much lower is that than the average life expectancy in the top 3 locations, to the nearest tenth?

   3.3 years

4. The Ralstons are driving from Washington, D.C., to Los Angeles. The trip is 4,234.185 kilometers. The Ralstons want to make the trip in 5 days, driving the same distance each day. About how far will they drive each day? Round your answer to the nearest whole number.

   about 847 km per day

Choose the letter of the best answer.

5. A top-grossing rock tour sold $121.2 million worth of tickets over the course of 60 shows. What was the average value of the tickets sold at each of the shows?

   A $20.2 million
   Ⓑ $2.02 million
   C $0.202 million
   D $7.27 million

6. A pizza parlor sells an 8-slice pizza for $9.20. If the pizza parlor charges a total of $1.20 more for a pizza it sells by the slice than for a pizza pie it sells whole, how much does a slice cost?

   F $1.15
   G $1.25
   Ⓗ $1.30
   J $1.50

7. Bill orders 3 copies of a book from a book club. The cost is $8.00, which includes $0.50 for handling. How much does the book cost?

   A $2.66   C $0.25
   Ⓑ $2.50   D $4.00

8. An apple pie has 27.6 grams of fat. It is cut into 6 slices. How many grams of fat are in 2 slices?

   F 4.6 grams   H 13.8 grams
   Ⓖ 9.2 grams   J 92.0 grams

Holt Mathematics

Reading Strategies
Use a Flowchart

Dividing a decimal by an integer is like dividing by a whole number.

Step 1: Place a decimal point in the quotient above the decimal point in the dividend.

Step 2: Divide as you would with whole numbers.

Step 3: Estimate to check to see if the answer is reasonable.

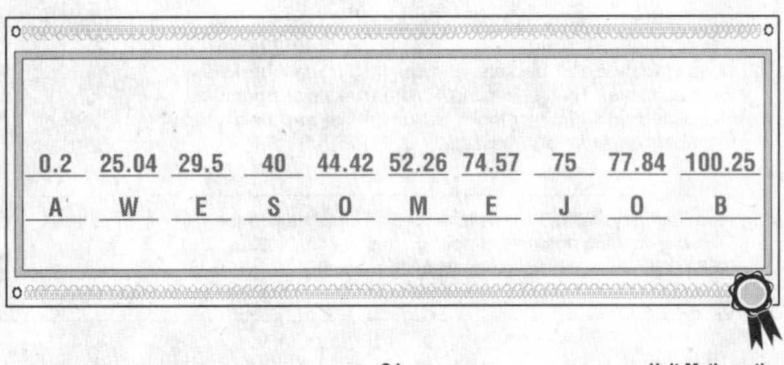

$35 \div 7 = 5$

Use the flowchart and the problem above to answer each question.

1. Where should you place the decimal point in the quotient?

   above the decimal point in the dividend

2. Do you think the estimate in the problem above is reasonable? Why or why not?

   Possible answer: yes, because 5.12 is close to 5

Use the flowchart and the problem $48.5 \div -5$ to answer each question.

3. Show where the decimal point should be placed in this problem.

   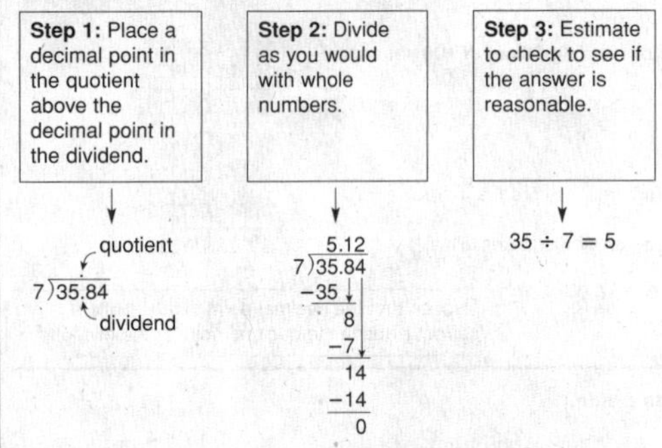

4. Divide. What is the answer?

   −9.7

5. Is the answer reasonable? Why or why not?

   −9.7 is a reasonable answer. $50 \div -5 = -10$; −9.7 is close to −10.

Holt Mathematics

Puzzles, Twisters & Teasers
R. U. Serious?

For each equation shown, decide whether the given solution is reasonable or unreasonable and circle your answer. Then use the Secret Decoder to solve the riddle. You'll need to choose one letter not to use.

1. $42.98 \div 7 = 600.14$
   reasonable   (unreasonable)

2. $-94.72 \div 37 = -2.56$
   (reasonable)   unreasonable

3. $12.8 \div 4 = 32$
   reasonable   (unreasonable)

4. $21.47 \div 19 = 1.13$
   (reasonable)   unreasonable

5. $0.148 \div 4 = .0037$
   reasonable   (unreasonable)

6. $65.28 \div 32 = 2.04$
   (reasonable)   unreasonable

7. $0.84 \div 12 = 7.12$
   reasonable   (unreasonable)

8. $20.95 \div 5 = 4.19$
   (reasonable)   unreasonable

9. $-39.2 \div 14 = 2.8$
   reasonable   (unreasonable)

10. $14.58 \div 3 = 4.86$
    (reasonable)   unreasonable

Secret Decoder

Reasonable	Unreasonable
D J	M B
N A O	P S E

Who is the most famous underwater spy in the world?

J	A	M	E	S
R	R	U	U	U

P	O	N	D
U	R	R	R

Holt Mathematics

Holt Mathematics

Practice A
Dividing Decimals and Integers by Decimals

Divide. Estimate to check whether your answer is reasonable.

1. $7.5\overline{)15}$ — 2

2. $1.2\overline{)72}$ — 60

3. $1.5\overline{)45}$ — 30

4. $7.5\overline{)-22.5}$ — -3

5. $4.8\overline{)16.8}$ — 3.5

6. $-2.7\overline{)11.07}$ — -4.1

Divide. Estimate to check whether your answer is reasonable.

7. $2.8\overline{)14}$ — 5

8. $-5.6\overline{)21}$ — -3.75

9. $3.2\overline{)48}$ — 15

10. $2.25\overline{)9}$ — 4

11. $2.4\overline{)6}$ — 2.5

12. $-1.25\overline{)65}$ — -52

13. Jessie used 2.7 gallons of gas to drive her car 72.9 miles. What was her car's gas mileage? — 27 miles per gallon

14. Ernesto bicycled 267 miles last week at an average speed of 8.9 mi/h. How many hours did he bicycle? — 30 hours

Holt Mathematics

Practice B
Dividing Decimals and Integers by Decimals

Divide.

1. $6 \div 0.25$ — 24

2. $78.74 \div 12.7$ — 6.2

3. $734.8 \div -1.67$ — -440

4. $56.525 \div 0.85$ — 66.5

5. $44.22 \div (-6.7)$ — -6.6

6. $-6.46 \div 0.04$ — -161.5

Divide. Estimate to check whether your answer is reasonable.

7. $63 \div (-4.5)$ — -14

8. $8 \div 3.2$ — 2.5

9. $87 \div 7.25$ — 12

10. $-36 \div 1.6$ — 22.5

11. $42 \div 4.8$ — 8.75

12. $90 \div 0.36$ — 250

13. Freddie used 6.75 gallons of gas to drive 155.25 miles. What was his car's gas mileage? — 23 miles per gallon

14. The members of a book club met at a restaurant for dinner. The total bill was $112.95 and they shared the bill equally. Each person paid $12.55. How many members are there in the book club? — 9 members

Holt Mathematics

Practice C
Dividing Decimals and Integers by Decimals

Divide. Estimate to check whether each answer is reasonable.

1. $39.984 \div (-5.1)$

2. $355.94 \div 4.81$

3. $-96 \div 1.28$

4. $34.146 \div 6.3$ — -7.84

5. $21.384 \div (-2.4)$ — 74

6. $15 \div 0.4$ — -75

7. $338.742 \div 0.54$ — 5.42

8. $401.598 \div (-0.81)$ — -8.91

9. $166.17 \div (-0.29)$ — 37.5

Row of answers: 627.3, -495.8, -573

Simplify each expression.

10. $40 \cdot (1.8 \div 0.9) \cdot 0.25$ — 20

11. $(32.4 \div 6.75) - 2.9 - 4.8$ — -2.9

12. $4.4 + (2.7 - 6.5) \div 0.4 + 0.6$ — -4.5

13. $6.25 \cdot 3.6 \div 1.25 \div 3$ — 6

14. $2.3 + (4.2 \cdot 1.32) \cdot 5 + 0.7$ — 30.72

15. $(-7.2 \cdot 0.9) + 3.9 + 6.48$ — 3.9

16. The price of binders at the school store increases by $0.25 each year. How long will it take for the price to increase by $3.00? — 12 years

Holt Mathematics

Reteach
Dividing Decimals and Integers by Decimals

To divide a decimal by a decimal:

Step 1: Make the divisor a whole number by moving the decimal point to the right.

Step 2: Move the decimal point in the dividend the same number of places. Remember to place the decimal in the quotient directly above the decimal point in the dividend.

Step 3: Divide.

Divide: $1.68 \div 0.3 \rightarrow 0.3\overline{)1.68} \rightarrow 3\overline{)16.8}$

$$3\overline{)16.8} = 5.6$$
-15
18
-18

Complete.

1. $5.6\overline{)4.48}$ — 0.8

a. How many decimal places are in the divisor? — 1

b. How many places do you need to move each decimal point? — 1

c. Rewrite the division. — $56\overline{)44.8}$

d. Complete the division. What is the quotient? — 0.8

Divide.

2. $5.2\overline{)3.64}$ — 0.7

3. $0.09\overline{)36.45}$ — 405

4. $0.59\overline{)0.708}$ — 1.2

Holt Mathematics

Holt Mathematics

Reteach
Dividing Decimals and Integers by Decimals (continued)

Sometimes it is necessary to write zeros in the dividend.

$$6 \div 0.25 \rightarrow 0.25\overline{)6} \rightarrow 0.25\overline{)6.00} \rightarrow 25\overline{)600}$$

$$\begin{array}{r} 24 \\ 25\overline{)600} \\ -50 \\ \hline 100 \\ -100 \\ \hline \end{array}$$

The divisor, 0.25, has 2 decimal places. In order to move the decimal point 2 places to the right in the divisor and the dividend, you need to write 2 zeros in the dividend.

Complete.

5. $0.35\overline{)7}$ quotient: **20**

a. How many decimal places are in the divisor? **2**

b. How many places do you need to move each decimal point? **2**

c. How many zeros do you need to write in the dividend? **2**

d. Complete the division. What is the quotient? **20**

Divide.

6. $1.6\overline{)8}$ **5**

7. $0.12\overline{)19.2}$ **160**

8. $1.25\overline{)48}$ **38.4**

39   **Holt Mathematics**

Challenge
Go For the Gold

A **golden rectangle** is a rectangle with a special relationship between the length and the width. The ancient Greeks believed that the shape of the golden rectangle was especially pleasing to the eye.

Discover the relationship between the length and width that makes a rectangle a golden rectangle.

1. The length of a golden rectangle is 5.26 centimeters and the width is 3.25 centimeters. Use a centimeter ruler to draw the rectangle.

Check students' drawings; rectangle should be about 5.3 cm long and 3.3 cm wide.

2. The **golden ratio** is a value that characterizes the relationship between the length and width of a golden rectangle. Use the dimensions of the golden rectangle above. Divide the length by the width to find the golden ratio. Round to the nearest hundredth.

1.62

For each length given below, use the golden ratio to find the width to make a golden rectangle. Round your answers to the nearest hundredth.

3. length = 8 meters **4.94 meters**

4. length = 4.85 inches **2.99 inches**

5. length = 6.09 feet **3.76 feet**

6. length = 12 yards **7.41 yards**

7. length = 16.5 centimeters **10.19 centimeters**

8. length = 24.6 meters **15.19 meters**

9. length = 55 inches **33.95 inches**

10. length = 83.2 feet **51.36 feet**

11. length = 98.2 yards **60.62 yards**

12. length = 116 centimeters **71.60 centimeters**

13. length = 220.5 inches **136.11 inches**

14. length = 348.7 meters **215.25 meters**

40   **Holt Mathematics**

Problem Solving
Dividing Decimals and Integers by Decimals

Write the correct answer.

1. Fran has a plank of wood 4.65 meters long. She wants to cut it into pieces 0.85 meters long. How many pieces of wood that length can she cut from the plank?

5 pieces

2. Jeremy has a box of 500 nails that weighs 1.35 kilograms. He uses 60 nails to build a birdhouse. How much do the nails in the birdhouse weigh?

0.162 kilogram

3. Rhosanda is downloading a file from the Internet. The size of the file is 7.45 MB. The file is downloading at the rate of 0.095 MB per second. How many seconds will it take to download the entire file? Round your answer to the nearest second.

79 seconds

4. Sean has a piece of poster board with an area of 476.28 square centimeters. He cuts it into equal-sized squares, each with an area of 39.69 square centimeters. How many squares can he cut from the piece of poster board?

12 squares

Choose the letter for the best answer.

5. Mount McKinley, in Alaska, is about 3.848 miles high. If a mountain climber can climb 0.25 miles per day, about how long, to the nearest day, would it take to climb Mount McKinley?

A about 5 days
B about 8 days
C about 12 days
D about 15 days

6. In 2000, a production worker in Japan who worked 38.5 hours in a week would have earned an average of $847. What was the hourly wage?

F $2.20 per hour
G $3.20 per hour
H $22.00 per hour
J $32.00 per hour

7. In the 2000 Summer Olympics, Michael Johnson ran the 400-meter race in 43.84 seconds. To the nearest hundredth, what was his speed in meters per second?

A 9.12 meters per second
B 10.96 meters per second
C 1.09 meters per second
D 91.24 meters per second

8. In the 2000 Summer Olympics, the United States relay team ran the 1,600-meter relay in 2 minutes, 56.35 seconds. To the nearest hundredth, what was the speed for the relay race in meters per second?

F 2.83 meters per second
G 6.24 meters per second
H 9.07 meters per second
J 28.39 meters per second

41   **Holt Mathematics**

Reading Strategies
Use a Visual Model

John has a piece of lumber 1.5 meters long. He needs to cut it into pieces that are 0.3 meter long. How many pieces can he cut? The number line shows a model of the problem.

$$1.5 \div 0.3 = 5$$

Sarah has 15 feet of yarn. She needs to cut it into lengths of 3 feet each. How many pieces can she cut? The number line shows a model of the problem.

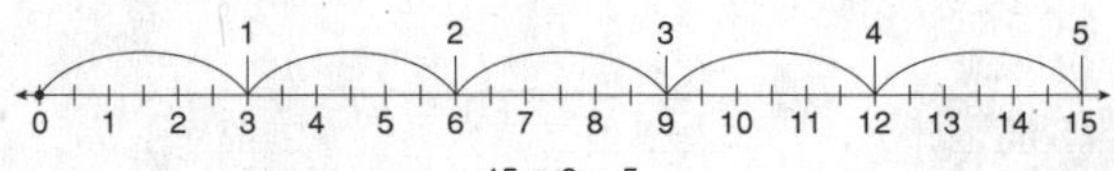

$$15 \div 3 = 5$$

Answer each question.

1. Compare the equations for the number lines above. What is the same about the equations?

Possible answer: The numerals are the same.

2. What is different?

Possible answer: One equation uses decimal values. The other uses whole numbers.

3. Compare the quotients of both problems. What do you notice?

They are both 5.

4. How can you change 1.5 to 15?

Move the decimal point one place to the right.

5. How can you change 0.3 to 3?

Move the decimal point one place to the right.

6. If you moved the decimal point in **both** the divisor and the dividend, would the quotient change?

no

42   **Holt Mathematics**

Holt Mathematics

Puzzles, Twisters & Teasers
By the Book!

Match the equations with their answers by writing the letter of the correct answers on the lines. When you are finished, read the letters from left to right and top to bottom as you would any book. If your answers are correct, the letters will spell out the title of a book. Write the book title on the blanks below.

32.1 **E**	0.9 **D**	−15 **N**	−5.3 **O**
5 **O**	6.25 **Y**	−0.5 **H**	120 **O**
0.4 **P**	309 **U**	0.046 **A**	14 **H**
5.6 **R**	−16 **E**	4 **A**	

$3.78 \div 4.2 = \underline{D}$ $1.06 \div -0.2 = \underline{O}$ $7.5 \div 1.2 = \underline{Y}$

$24 \div 0.2 = \underline{O}$ $9.27 \div 0.03 = \underline{U}$ $1.12 \div 0.08 = \underline{H}$

$-40 \div 2.5 = \underline{E}$ $5 \div 1.25 = \underline{A}$ $14 \div 2.5 = \underline{R}$

$1.15 \div 25 = \underline{A}$ $4.12 \div 10.3 = \underline{P}$ $-1.6 \div 3.2 = \underline{H}$

$12.0 \div 2.4 = \underline{O}$ $36 \div -2.4 = \underline{N}$ $96.3 \div 3 = \underline{E}$

$\underline{D} \quad \underline{O} \quad \underline{Y} \quad \underline{O} \quad \underline{U} \quad \underline{H} \quad \underline{E} \quad \underline{A} \quad \underline{R}$

$\underline{A} \quad \underline{P} \quad \underline{H} \quad \underline{O} \quad \underline{N} \quad \underline{E} ?$

Now circle the name of someone who might have been the author of this book.

By:

Isabel Ringing   Justin Case

M. T. Wallet   Hugo First

43   **Holt Mathematics**

Practice A
Solving Equations Containing Decimals

Solve. Choose the letter for the best answer.

1. $t + 0.7 = 9$
   A $t = 9.7$ C $t = 6.3$
   (B) $t = 8.3$ D $t = 0.63$

2. $p - 1.6 = 11$
   F $p = 6.875$ (H) $p = 12.6$
   G $p = 9.4$ J $p = 17.6$

3. $\frac{h}{3} = 1.5$
   A $h = 0.5$ C $h = 9$
   (B) $h = 4.5$ D $h = 45$

4. $7z = 2.1$
   F $z = -4.9$ H $z = 14.7$
   (G) $z = 0.3$ J $z = 2.8$

Solve.

5. $x - 5.1 = 4.8$ 6. $h + 6.9 = 12.7$ 7. $k + 9.2 = -7.6$

$x = 9.9$ $h = 5.8$ $k = -16.8$

8. $g - 4.44 = 2.4$ 9. $0.18 + w = 0.75$ 10. $m - 3.1 = 9.65$

$g = 6.84$ $w = 0.57$ $m = 12.75$

11. $4.2n = 14.7$ 12. $9.7j = 58.2$ 13. $5.6p = -11.76$

$n = 3.5$ $j = 6$ $p = -2.1$

14. $43.2 = 2.7y$ 15. $64.6 = 6.8x$ 16. $40.32 = 12.6m$

$y = 16$ $x = 9.5$ $m = 3.2$

17. $\frac{s}{5.4} = 6$ 18. $\frac{f}{0.8} = 7$ 19. $\frac{d}{4.6} = 0.7$

$s = 32.4$ $f = 5.6$ $d = 3.22$

20. $\frac{c}{0.4} = 1.75$ 21. $\frac{h}{6.1} = 12$ 22. $\frac{a}{0.35} = 8.4$

$c = 0.7$ $h = 73.2$ $a = 2.94$

23. At the movie theater, a gift certificate for 15 tickets costs $93.75. What is the cost of each ticket? $6.25

24. Four couples are going to a concert. If each ticket costs $5.90, how much will it cost to buy tickets for all 8 people? $47.20

44   **Holt Mathematics**

Practice B
Solving Equations Containing Decimals

Solve.

1. $t + 0.77 = 9.3$ 2. $p - 1.34 = -11.8$ 3. $r + 2.14 = 7.8$

$t = 8.53$ $p = -10.46$ $r = 5.66$

4. $3.65 + e = -1.4$ 5. $w - 16.7 = 8.27$ 6. $z - 17.2 = 7.13$

$e = -5.05$ $w = 24.97$ $z = 24.33$

7. $p - 67.5 = 24.81$ 8. $h + 26.9 = 12.74$ 9. $k + 89.2 = -47.62$

$p = 92.31$ $h = -14.16$ $k = -136.82$

10. $x - 0.45 = 5.97$ 11. $1.08 + n = 15.72$ 12. $y - 6.32 = 0.73$

$x = 6.42$ $n = 14.64$ $y = 7.05$

13. $4.3p = 28.81$ 14. $7.7j = 76.23$ 15. $3.8g = -104.12$

$p = 6.7$ $j = 9.9$ $g = -27.4$

16. $18.36 = 2.7y$ 17. $99.96 = 6.8x$ 18. $293.92 = 17.6c$

$y = 6.8$ $x = 14.7$ $c = 16.7$

19. $\frac{e}{7.4} = 6.9$ 20. $\frac{f}{12.7} = 15.6$ 21. $\frac{d}{9.7} = 20.8$

$e = 51.06$ $f = 198.12$ $d = 201.76$

22. $\frac{w}{-0.2} = 15.4$ 23. $\frac{m}{9.8} = 1.7$ 24. $\frac{s}{14.35} = -5.2$

$w = -3.08$ $m = 16.66$ $s = -74.62$

25. Jeff paid a flat fee of $269.50 for a year's worth of vet visits for his 4 cats. He made 14 visits during the year. What was the average cost per visit?

$19.25

45   **Holt Mathematics**

Practice C
Solving Equations Containing Decimals

Solve.

1. $t + 4.77 = 9.38$ 2. $p - 1.34 = -17.08$ 3. $2.55c = 19.89$

$t = 4.61$ $p = -15.74$ $c = 7.8$

4. $3.65 + e = -10.74$ 5. $16.17 - w = 8.27$ 6. $34.932 = 4.92g$

$e = -14.39$ $w = 7.9$ $g = 7.1$

7. $67.75 - a = 34.81$ 8. $h + 26.99 = 11.74$ 9. $k + 89.72 = -47.62$

$a = 32.94$ $h = -15.25$ $k = -137.34$

10. $r + 2.14 = 27.84$ 11. $6.45j = 36.765$ 12. $2.91p = -24.444$

$r = 25.7$ $j = 5.7$ $p = -8.4$

13. $29.406 = 7.54y$ 14. $41.096 = 4.67x$ 15. $17.52 - z = 4.13$

$y = 3.9$ $x = 8.8$ $z = 13.39$

16. $\frac{a}{7.4} = 6.79$ 17. $55.7 + b = 32.119$ 18. $\frac{d}{9.7} = 20.78$

$a = 50.246$ $b = -23.581$ $d = 201.566$

19. $0.378 + s = 2.918$ 20. $v - 4.025 = 1.9024$ 21. $\frac{f}{12.7} = 15.26$

$s = 2.54$ $v = 5.9274$ $f = 193.802$

22. $\frac{n}{0.38} = 5.89$ 23. $\frac{t}{-1.88} = -6.2$ 24. $\frac{w}{0.0056} = -4.7$

$n = 2.2382$ $t = 11.656$ $w = -0.02632$

25. Sam lives 3.5 times farther from school than Sara does. Sam lives 13.3 miles from school. How far from school does Sara live?   3.8 miles

26. The Bow-Wow Pet Shop made a profit of $74,239 this year. This was $81,668 more than the profit from last year. What was the profit last year?   −$7,429

46   **Holt Mathematics**

---

          113   **Holt Mathematics**

Reteach
Solving Equations Containing Decimals

You can solve equations with decimals the same way you solve equations with whole numbers. Remember to always perform the same calculation on both sides of the equation to keep the two sides equal.

- You can use addition to solve a subtraction equation involving decimals.

$$x - 1.45 = 6.7$$
$$x - 1.45 + \mathbf{1.45} = 6.7 + \mathbf{1.45}$$
$$x = 8.15$$

Addition undoes subtraction.

- You can use subtraction to solve an addition equation involving decimals.

$$n + 24.8 = -15.2$$
$$n + 24.8 - \mathbf{24.8} = -15.2 - \mathbf{24.8}$$
$$n = -40$$

Subtraction undoes addition.

Solve.

1. $e + 7.1 = 9.3$
$e + 7.1 - 7.1 = 9.3 - 7.1$
$e = \underline{2.2}$

2. $x - 1.9 = 5.4$
$x - 1.9 + \underline{1.9} = 5.4 + \underline{1.9}$
$x = \underline{7.3}$

3. $w - 8.3 = -4.12$
$w - 8.3 + \underline{8.3} = -4.12 + \underline{8.3}$
$w = \underline{4.18}$

4. $b + 5.75 = -6.2$
$b + 5.75 - \underline{5.75} = -6.2 - \underline{5.75}$
$b = \underline{-11.95}$

5. $t + 39.5 = 54.1$
$t = 14.6$

6. $p - 29.4 = 3.7$
$p = 33.1$

7. $r - 6.25 = -17.3$
$r = -11.05$

8. $k + 9.8 = -11.9$
$k = -21.7$

Holt Mathematics

Reteach
Solving Equations Containing Decimals (continued)

- You can use division to solve a multiplication equation involving decimals.

$$3.6y = 9$$
$$\frac{3.6y}{3.6} = \frac{9}{3.6}$$
$$y = 2.5$$

Division undoes multiplication.

- You can use multiplication to solve a division equation involving decimals.

$$\frac{a}{4.2} = 18$$
$$4.2 \cdot \frac{a}{4.2} = 18 \cdot 4.2$$
$$a = 75.6$$

Multiplication undoes division.

Solve.

9. $5.7g = 45.6$
$5.7g \div \underline{5.7} = 45.6 \div \underline{5.7}$
$g = \underline{8}$

10. $-6f = 8.04$
$-6f \div \underline{-6} = 8.04 \div \underline{-6}$
$f = \underline{-1.34}$

11. $\frac{n}{0.14} = 15$
$\underline{0.14} \cdot \frac{n}{0.14} = 15 \cdot \underline{0.14}$
$n = \underline{2.1}$

12. $\frac{m}{6.3} = -9.1$
$\underline{6.3} \cdot \frac{m}{6.3} = -9.1 \cdot \underline{6.3}$
$m = \underline{-57.33}$

13. $8y = 93.6$
$y = 11.7$

14. $-3.4c = 20.74$
$c = -6.1$

15. $\frac{s}{10.5} = 3.8$
$s = 39.9$

16. $\frac{h}{0.4} = -7.2$
$h = -2.88$

Holt Mathematics

Challenge
The Pairs Event

When you solve a system of equations, you solve two different equations at once. The solution to a system of equations is an ordered pair (x, y).

Example: $2.5x = 50$
$x + y = 10.5$

First, solve the first equation for x: $x = 20$.

Then use the solution from the first equation to solve for y in the second equation.

$x = 20$

$x + y = 10.5$
$20 + y = 10.5$
$y = -9.5$

The solution to the system of equations is the ordered pair $(20, -9.5)$.

Solve each system of equations.

1. $x - 5.4 = 7.3$
$2x + y = 10$
$(12.7, -15.4)$

2. $1.4x = 2.8$
$\frac{y}{x} = 9.6$
$(2, 19.2)$

3. $\frac{x}{3.9} = 0.8$
$4y = x$
$(3.12, 0.78)$

4. $x + 9.25 = 10.1$
$y - 3x = 1.7$
$(0.85, 4.25)$

5. $6.4x = 2.56$
$x = 2y$
$(0.4, 0.2)$

6. $x + 3.02 = 5$
$\frac{y}{3.5} = x$
$(1.98, 6.93)$

7. $6x = -7.2$
$y - x = 3.4$
$(-1.2, 2.2)$

8. $\frac{x}{0.3} = -9$
$y + x = 0.6$
$(-2.7, 3.3)$

9. $x - 4.7 = -9$
$\frac{y}{x} = 0.5$
$(-4.3, -2.15)$

10. $x + 8.4 = -2$
$2x + y = 1.1$
$(-10.4, 21.9)$

11. $2.6x = 3.64$
$-3x = y$
$(1.4, -4.2)$

12. $\frac{x}{0.4} = -3.2$
$x + y = 0.35$
$(-1.28, 1.63)$

Holt Mathematics

Problem Solving
Solving Equations Containing Decimals

Write the correct answer.

1. The diameter of the secondary mirror in NASA's Hubble telescope is 0.3 meter. The primary mirror is 8 times as large. What is the diameter of the primary mirror?

2.4 meters

2. A cubic centimeter of titanium weighs 4.507 grams. The same volume of gold weighs 19.3 grams. How much more does a cubic centimeter of gold weigh?

14.793 grams

3. Brianna drives 3.35 miles to work every day. This is 1.75 miles less than Darius drives to work every day. How far does Darius drive to work?

5.1 miles

4. The weight of an object on Mars is 0.38 times its weight on Earth. How much would a 125-pound person weigh on Mars?

47.5 pounds

Choose the letter for the best answer.

The table shows the orbital velocity of some of the planets.

5. How much greater is the orbital velocity of Mercury than that of Uranus?

A 33.97 miles per second
B 25.51 miles per second
C 7.03 miles per second
D 4.23 miles per second

6. How many miles does Mercury travel in an hour?

F 1,784.4 miles
G 107,064 miles
H 10,706.4 miles
J 17,844 miles

7. How many miles does Jupiter travel in a minute?

A 8.12 miles
B 81.2 miles
C 48.72 miles
D 487.2 miles

Planets' Orbital Velocity	
Planet	**Orbital Velocity (mi/s)**
Mercury	29.74
Venus	21.76
Earth	18.5
Mars	14.99
Jupiter	8.12
Saturn	6.02
Uranus	4.23

8. During the time it takes Saturn to travel 32,508 miles, how much time has elapsed on Earth?

F 5,400 minutes
G 1,757.19 minutes
H 195,698 seconds
J 90 minutes

Holt Mathematics

Holt Mathematics

Reading Strategies
3-6 Compare and Contrast

Compare the steps for solving equations with whole numbers to the steps for solving equations with decimals.

Solving Equations with Whole Numbers	Example:
Step 1: This is a subtraction problem. Add to get x by itself.	$x - 145 = 1{,}720$
Step 2: Add **145** to both sides of the equation.	$x - 145 + 145 = 1{,}720$ **+ 145**
Step 3: Solve.	$x = 1{,}865$

Solving Equations with Decimals	Example:
Step 1: This is a subtraction problem. Add to get x by itself.	$x - 1.45 = 17.2$
Step 2: Add **1.45** to both sides of the equation.	$x - 1.45 + 1.45 = 17.2$ **+ 1.45**
Step 3: Solve.	$x = 18.65$

Use the chart to answer the following questions.

1. Compare the steps in solving an equation with whole numbers to the steps for an equation with decimals. What do you notice?

   __The steps are the same.__

2. What is different about solving an equation with whole numbers and solving an equation with decimals?

   __Whole numbers are used in whole number equations and decimals are used in decimal equations.__

Compare solving a multiplication equation with whole numbers to one with decimals: $3y = 702$; $3y = 7.02$. Answer each question.

3. What is the first step in solving both equations?

   __Get y by itself on one side of the equation.__

4. What operation will you use first in the two equations?

   __division__

5. Compare the number you divide by on both sides of the whole number equation to the number you divide by in the decimal equation.

   __Divide by 3 in both types of equations.__

51

Puzzles, Twisters & Teasers
3-6 Messy Problems!

Solve the equations. Then use the variables to answer the riddle.

$-1.8 + g = -3.8$	$g = \underline{-2}$
$60t = 54$	$t = \underline{0.9}$
$1.05 = -7n$	$n = \underline{-0.15}$
$h + 0.48 = 1.2$	$h = \underline{0.72}$
$7.9 = i + 12.7$	$i = \underline{-4.8}$
$e + 0.81 = -6.3$	$e = \underline{-7.11}$
$\frac{l}{0.8} = 4.9$	$l = \underline{3.92}$
$y - 4.1 = -5$	$y = \underline{-0.9}$
$1.2b = -1.44$	$b = \underline{-1.2}$
$\frac{a}{2.4} = 2.7$	$a = \underline{6.48}$
$72 = 4.5r$	$r = \underline{16}$
$3.2w = 8$	$w = \underline{2.5}$
$\frac{d}{-6.4} = 0.85$	$d = \underline{-5.44}$
$s - 9.01 = 12.6$	$s = \underline{21.61}$
$x + 30.34 = -22.87$	$x = \underline{-53.21}$

Why are basketball players sloppy eaters?

T	H	E	Y	'	R	E
0.9	0.72	−7.11	−0.9		16	−7.11

A	L	W	A	Y	S
6.48	3.92	2.5	6.48	−0.9	21.61

D	R	I	B	B	L	I	N	G
−5.44	16	−4.8	−1.2	−1.2	3.92	−4.8	−0.15	−2

52

Practice A
3-7 Estimate with Fractions

Estimate each sum or difference.

1. $\frac{1}{6} + \frac{5}{8}$

   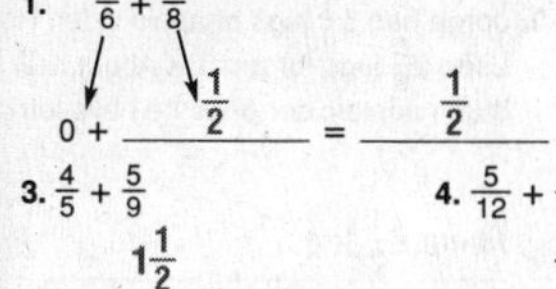
   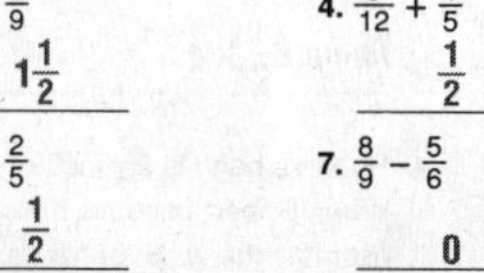

   $0 + \frac{1}{2} = \frac{1}{2}$

2. $4\frac{7}{8} - 2\frac{1}{10}$

   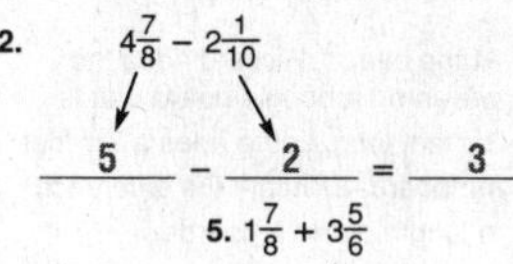

   $5 - 2 = 3$

3. $\frac{4}{5} + \frac{5}{9}$

   $1\frac{1}{2}$

4. $\frac{5}{12} + \frac{1}{5}$

   $\frac{1}{2}$

5. $1\frac{7}{8} + 3\frac{5}{6}$

   6

6. $\frac{7}{8} - \frac{2}{5}$

   $\frac{1}{2}$

7. $\frac{8}{9} - \frac{5}{6}$

   0

8. $8\frac{3}{5} - 3\frac{9}{10}$

   $4\frac{1}{2}$

Estimate each product or quotient.

9. $8\frac{1}{8} \cdot 9\frac{8}{9}$

   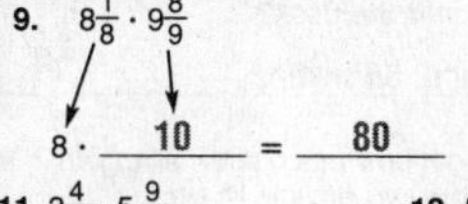

   $8 \cdot 10 = 80$

10. $39\frac{4}{5} \div 7\frac{9}{10}$

    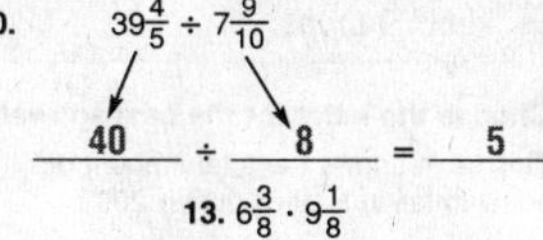

    $40 \div 8 = 5$

11. $3\frac{4}{5} \cdot 5\frac{9}{10}$

    24

12. $5\frac{1}{5} \cdot 3\frac{5}{6}$

    20

13. $6\frac{3}{8} \cdot 9\frac{1}{8}$

    54

14. $23\frac{5}{8} \div 6\frac{1}{5}$

    4

15. $29\frac{7}{9} \div 9\frac{3}{4}$

    3

16. $9\frac{1}{5} \div 2\frac{7}{8}$

    3

17. Erin hikes $3\frac{3}{4}$ miles in the morning and $5\frac{1}{8}$ miles in the afternoon. Estimate the number of miles Erin hikes in all.

    __9 miles__

18. A trail is $16\frac{3}{8}$ miles long. Max has hiked $4\frac{13}{16}$ miles along the trail so far. Estimate the number of miles Max has left to hike.

    __$11\frac{1}{2}$ miles__

53

Practice B
3-7 Estimate with Fractions

Estimate each sum or difference.

1. $\frac{5}{11} + \frac{4}{9}$

   1

2. $\frac{6}{13} + \frac{8}{9}$

   $1\frac{1}{2}$

3. $\frac{9}{10} - \frac{4}{9}$

   $\frac{1}{2}$

4. $1\frac{5}{8} - \frac{4}{7}$

   1

5. $3\frac{7}{8} + \left(-\frac{2}{5}\right)$

   $3\frac{1}{2}$

6. $\frac{8}{9} - \frac{1}{12}$

   1

7. $4\frac{5}{16} + 2\frac{9}{10}$

   $7\frac{1}{2}$

8. $11\frac{3}{7} - 5\frac{5}{6}$

   $5\frac{1}{2}$

9. $7\frac{1}{16} + \left(-\frac{11}{12}\right)$

   6

Estimate each product or quotient.

10. $12\frac{2}{5} \div 5\frac{3}{4}$

    2

11. $7\frac{7}{8} \cdot 4\frac{3}{5}$

    40

12. $5\frac{1}{6} \cdot 3\frac{2}{9}$

    15

13. $23\frac{7}{10} \div 4\frac{2}{5}$

    6

14. $17\frac{11}{12} \div 8\frac{5}{9}$

    2

15. $8\frac{7}{12} \cdot 6\frac{9}{10}$

    63

16. $12\frac{3}{8} \cdot 6\frac{1}{6}$

    72

17. $35\frac{2}{3} \div 3\frac{5}{7}$

    9

18. $16\frac{5}{8} \cdot 2\frac{1}{5}$

    34

19. A hallway has a length of $15\frac{3}{4}$ feet and a width of $4\frac{1}{12}$ feet. Estimate the area of the hallway in square feet.

    __64 square feet__

20. A 6-week old puppy weighed $8\frac{7}{16}$ pounds. At 12 weeks of age, the same puppy weighed about $17\frac{3}{8}$ pounds. Estimate how much weight the puppy gained between the ages of 6 weeks and 12 weeks.

    __9 pounds__

54

LESSON 3-7
Practice C
Estimate with Fractions

Estimate each sum or difference.

1. $\frac{6}{11} + \frac{6}{7}$

$1\frac{1}{2}$

2. $\frac{5}{12} - \frac{2}{5}$

0

3. $4\frac{4}{9} - 3\frac{1}{8}$

$1\frac{1}{2}$

4. $5\frac{4}{7} + \frac{7}{12}$

6

5. $5\frac{7}{8} + \left(-2\frac{1}{16}\right)$

4

6. $6\frac{9}{10} - 2\frac{7}{16}$

$4\frac{1}{2}$

7. $8\frac{11}{12} + 1\frac{5}{15} - 3\frac{6}{11}$

$6\frac{1}{2}$

8. $3\frac{7}{15} + 5\frac{6}{7} - 1\frac{1}{8}$

$8\frac{1}{2}$

9. $2\frac{3}{20} - 7\frac{7}{9} + 15\frac{9}{17}$

$9\frac{1}{2}$

Estimate each product or quotient.

10. $2\frac{3}{5} \cdot 12\frac{1}{4}$

36

11. $5\frac{7}{9} \div 2\frac{3}{10}$

3

12. $20\frac{2}{9} \div \left(-4\frac{5}{7}\right)$

-4

13. $7\frac{5}{16} \cdot \left(-3\frac{5}{8}\right)$

-28

14. $16\frac{7}{9} \cdot 3\frac{9}{20}$

51

15. $14\frac{11}{15} \div 5\frac{2}{9}$

3

16. $-6\frac{5}{18} \cdot \left(-7\frac{11}{12}\right)$

48

17. $47\frac{13}{16} \div 3\frac{2}{11}$

16

18. $-31\frac{4}{5} \div 7\frac{8}{15}$

4

19. Carlos weighs $125\frac{1}{4}$ pounds. Hector weighs $17\frac{3}{4}$ pounds less than Carlos. About how much does Hector weigh?

107 pounds

20. A storage closet is $3\frac{3}{4}$ feet wide, $4\frac{1}{6}$ feet long, and $7\frac{11}{12}$ feet high. Estimate the volume of the closet.

128 cubic feet

55

Holt Mathematics

LESSON 3-7
Reteach
Estimate with Fractions

You can estimate sums and differences of fractions and mixed numbers by rounding the fractions or mixed numbers to the nearest $\frac{1}{2}$. Use a number line to help.

Estimate $\frac{3}{8} + \frac{9}{10}$.

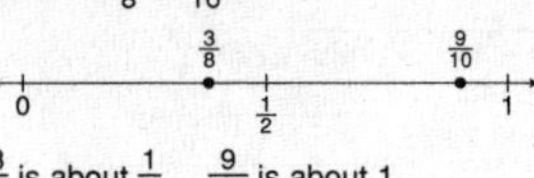

$\frac{3}{8}$ is about $\frac{1}{2}$. $\frac{9}{10}$ is about 1.

So, $\frac{3}{8} + \frac{9}{10}$ is about $\frac{1}{2} + 1$, or $1\frac{1}{2}$.

You can estimate products and quotients of mixed numbers by rounding the mixed numbers to the nearest whole numbers. Use a number line to help.

Estimate $2\frac{1}{4} \cdot 3\frac{5}{8}$.

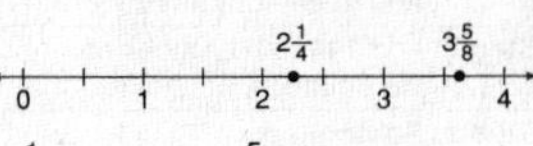

$2\frac{1}{4}$ is about 2. $3\frac{5}{8}$ is about 4.

So, $2\frac{1}{4} \cdot 3\frac{5}{8}$ is about $2 \cdot 4$, or 8.

Estimate.

1. $\frac{7}{8} - \frac{4}{7}$

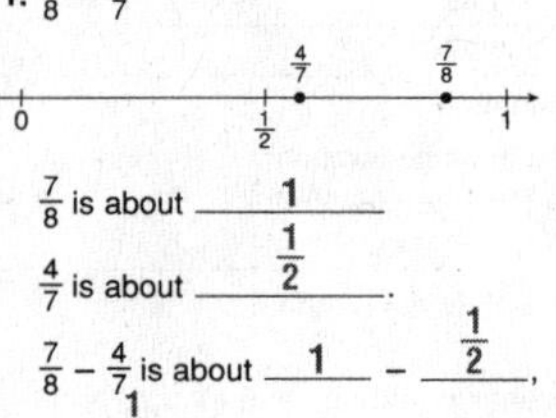

$\frac{7}{8}$ is about ___1___

$\frac{4}{7}$ is about ___$\frac{1}{2}$___.

$\frac{7}{8} - \frac{4}{7}$ is about ___1___ − ___$\frac{1}{2}$___,

or ___$\frac{1}{2}$___.

2. $4\frac{1}{6} \div 1\frac{5}{9}$

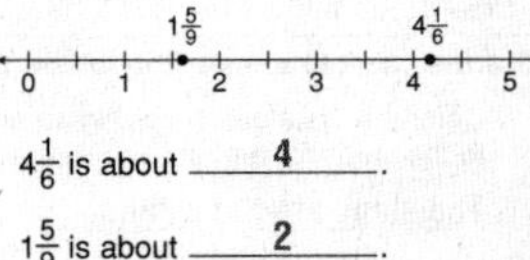

$4\frac{1}{6}$ is about ___4___.

$1\frac{5}{9}$ is about ___2___.

$4\frac{1}{6} \div 1\frac{5}{9}$ is about ___4___ ÷ ___2___,

or ___2___.

3. $\frac{11}{12} + \frac{2}{5}$

$1\frac{1}{2}$

4. $6\frac{5}{8} - 4\frac{5}{6}$

$1\frac{1}{2}$

5. $5\frac{8}{9} + 3\frac{7}{8}$

10

6. $3\frac{3}{8} \cdot 5\frac{7}{9}$

18

7. $23\frac{5}{6} \div 7\frac{7}{8}$

3

8. $5\frac{1}{4} \cdot 6\frac{2}{5}$

30

56

Holt Mathematics

LESSON 3-7
Challenge
Don't Underestimate the Answer to a Riddle!

Circle the fractions whose estimated sums, differences products, or quotients equal the given solution. To find the answer to each riddle, write the letters below the circled fractions in order in the blanks.

Why did the student go to school in an airplane?

1. $\frac{7}{8} + \left[1\frac{6}{11} \text{ or } 4\frac{4}{5}\right] - \left[3\frac{7}{8} \text{ or } \frac{9}{10}\right] \approx 3\frac{1}{2}$
   H   E   M   I

2. $5\frac{11}{12} + \left[6\frac{3}{8} \text{ or} \left(-4\frac{5}{6}\right)\right] \approx 1$
   P   G

3. $\left[\frac{6}{11} \text{ or } \frac{6}{7}\right] + \left[5\frac{9}{10} \text{ or } \frac{4}{19}\right] - \left[7\frac{2}{3} \text{ or } \left(-4\frac{16}{19}\right)\right] \approx 5\frac{1}{2}$
   H   T   Y   E   S   R

ANSWER: She wanted to get a ___HIGHER___ education!

What kind of hair does the Atlantic Ocean have?

4. $3\frac{1}{10} + \left[\frac{9}{10} \text{ or } 4\frac{5}{12}\right] - \left[-6\frac{1}{20} \text{ or } \left(-6\frac{9}{10}\right)\right] \approx 11$
   W   H   U   A

5. $8\frac{4}{7} - \left[\left(-6\frac{5}{9}\right) \text{ or } -6\frac{1}{8}\right] \approx 15$
   V   R

6. $\left[2\frac{1}{3} \text{ or } 1\frac{6}{13}\right] - 4\frac{1}{10} \approx -2$
   Y   T

ANSWER: ___WAVY___

What can go across the country and still stay in its corner?

7. $\left[4\frac{5}{8} \text{ or } 6\frac{1}{8}\right] \cdot \left[8\frac{3}{10} \text{ or } 12\frac{9}{10}\right] \approx 48$
   T   S   T   R

8. $\left[20\frac{1}{5} \text{ or } 29\frac{7}{8}\right] \div \left[6\frac{3}{5} \text{ or } 4\frac{11}{13}\right] \approx 6$
   U   A   C   M

9. $63\frac{2}{5} \div \left[8\frac{4}{5} \text{ or } 6\frac{7}{9}\right] \approx 7$
   P   K

ANSWER: A ___STAMP___

57

Holt Mathematics

LESSON 3-7
Problem Solving
Estimate with Fractions

Write the correct answer.

1. At the beach, Richard rides the waves on a boogie board that is $3\frac{2}{3}$ feet long. Laura rides a $7\frac{1}{2}$-foot surfboard. Estimate the difference in length of the 2 boards.

about 4 feet

2. Jorgé had $5\frac{1}{3}$ jugs of apple cider. He used $2\frac{5}{6}$ jugs for a party. About how much apple cider does he have left?

about $2\frac{1}{2}$ jugs

3. Sari jogs $2\frac{3}{4}$ miles on Monday, $3\frac{5}{6}$ miles on Wednesday, and $2\frac{1}{3}$ miles on Friday. Estimate her total distance for the week.

about 9 miles

4. Robert's hand is $2\frac{7}{8}$ inches wide. When Robert uses his hand to estimate the width of his desk, he finds that the desk is about $11\frac{3}{4}$ hands wide. About how many inches wide is the desk?

about 36 inches

Choose the letter for the best answer.

This table shows the total amount of snow to fall in 5 cities during 2003.

Snowfall During 2003

City	Amount of Snow (in.)
Chicago, IL	$17\frac{2}{5}$
Indianapolis, IN	$44\frac{1}{10}$
Marquette, MI	$191\frac{4}{5}$
Moline, IL	$23\frac{4}{5}$
Providence, RI	$58\frac{9}{10}$

5. About how much snow all together fell in the two cities in Illinois?
   A 40 inches   C 44 inches
   (B) 41 inches   D 61 inches

6. About how much more snow fell in Providence than in Indianapolis?
   F 35 inches   (H) 15 inches
   G 20 inches   J 10 inches

7. Which city had about 11 times as much snow as Chicago?
   A Indianapolis   C Moline
   (B) Marquette   D Providence

8. About how much more snow fell in Indianapolis than in Moline?
   (F) 20 inches   H 24 inches
   G 22 inches   J 30 inches

58

Holt Mathematics

Holt Mathematics

Reading Strategies
3-7 Analyze Information

You can use some easy rules to help you estimate fractions.

· If the numerator is much smaller than the denominator, round to 0.	**Fractions Close to 0** $\frac{1}{5}$ $\frac{2}{25}$ $\frac{6}{50}$
· If the numerator is about half the value of the denominator, round to $\frac{1}{2}$.	**Fractions Close to $\frac{1}{2}$** $\frac{6}{13}$ $\frac{9}{20}$ $\frac{12}{23}$
· If the numerator and denominator are close to the same value, round to 1.	**Fractions Close to 1** $\frac{7}{8}$ $\frac{23}{25}$ $\frac{57}{60}$

This can help you estimate sums and differences of fractions.

$$\frac{5}{9} - \frac{14}{16}$$
$$\downarrow \qquad \downarrow$$
$$\frac{1}{2} - 1 = -\frac{1}{2}$$

$$3\frac{3}{8} + 1\frac{8}{9}$$
$$\downarrow \qquad \downarrow$$
$$3\frac{1}{2} + 2 = 5\frac{1}{2}$$

Write *close to 0*, *close to $\frac{1}{2}$*, or *close to 1* for each fraction.

1. $\frac{13}{25}$ close to $\frac{1}{2}$

2. $\frac{9}{75}$ close to 0

3. $\frac{7}{9}$ close to 1

4. $\frac{14}{16}$ close to 1

5. $\frac{3}{20}$ close to 0

6. $\frac{1}{9}$ close to 0

7. $\frac{7}{15}$ close to $\frac{1}{2}$

59   **Holt Mathematics**

Puzzles, Twisters & Teasers
3-7 Math Makes Me Crabby!

Estimate the answers. Then solve the riddle.

H $35\frac{1}{3} \div 4\frac{9}{10}$ 7

T $2\frac{8}{9} + 1\frac{7}{8}$ 5

S $31\frac{7}{8} \div 3\frac{5}{9}$ 8

E $2\frac{2}{5} \cdot 2\frac{13}{15}$ 6

I $\frac{7}{9} - \frac{3}{8}$ $\frac{1}{2}$

Y $\frac{3}{5} + \frac{6}{7}$ $1\frac{1}{2}$

F $4\frac{5}{7} \cdot 5\frac{1}{3}$ 25

A $8\frac{3}{4} + -6\frac{1}{5}$ 3

L $15\frac{1}{7} + 10\frac{8}{9}$ 26

R $\frac{5}{6} + \frac{2}{12}$ 1

Why are crabs so greedy?

$$\underset{5}{T}\ \underset{7}{H}\ \underset{6}{E}\ \underset{1\frac{1}{2}}{Y} \qquad \underset{3}{A}\ \underset{1}{R}\ \underset{6}{E}$$

$$\underset{8}{S}\ \underset{7}{H}\ \underset{6}{E}\ \underset{26}{L}\ \underset{26}{L}\ \underset{25}{F}\ \underset{\frac{1}{2}}{I}\ \underset{8}{S}\ \underset{7}{H}.$$

60   **Holt Mathematics**

Practice A
3-8 Adding and Subtracting Fractions

Add or subtract. Write each answer in simplest form.

1. $\frac{1}{2} + \frac{1}{2}$ 1
2. $\frac{2}{5} + \frac{2}{5}$ $\frac{4}{5}$
3. $\frac{1}{4} + \frac{1}{4}$ $\frac{1}{2}$

4. $\frac{1}{8} + \frac{5}{8}$ $\frac{3}{4}$
5. $\frac{5}{6} + \frac{5}{6}$ $1\frac{2}{3}$
6. $\frac{7}{20} + \frac{17}{20}$ $1\frac{1}{5}$

7. $\frac{7}{12} - \frac{1}{12}$ $\frac{1}{2}$
8. $\frac{9}{16} - \frac{5}{16}$ $\frac{1}{4}$
9. $\frac{3}{10} - \frac{9}{10}$ $-\frac{3}{5}$

10. $\frac{7}{8} - \frac{5}{8}$ $\frac{1}{4}$
11. $\frac{4}{9} - \frac{7}{9}$ $-\frac{1}{3}$
12. $\frac{7}{15} - \frac{1}{15}$ $\frac{2}{5}$

13. $\frac{1}{3} + \frac{3}{4}$ $1\frac{1}{12}$
14. $\frac{1}{4} + \frac{5}{6}$ $1\frac{1}{12}$
15. $\frac{5}{8} + \frac{1}{6}$ $\frac{19}{24}$

16. $\frac{1}{4} - \frac{5}{6}$ $-\frac{7}{12}$
17. $\frac{3}{4} - \frac{1}{5}$ $\frac{11}{20}$
18. $\frac{9}{10} - \frac{3}{5}$ $\frac{3}{10}$

19. Michael bought $\frac{1}{2}$ pound of Swiss cheese. He used $\frac{1}{4}$ pound for a sandwich. How much was left?
$\frac{1}{4}$ pound

61   **Holt Mathematics**

Practice B
3-8 Adding and Subtracting Fractions

Add or subtract. Write each answer in simplest form.

1. $\frac{1}{5} + \frac{2}{5}$ $\frac{3}{5}$
2. $\frac{4}{15} + \frac{8}{15}$ $\frac{4}{5}$
3. $\frac{7}{12} - \frac{5}{12}$ $\frac{1}{6}$

4. $\frac{9}{10} - \frac{7}{10}$ $\frac{1}{5}$
5. $\frac{7}{12} - \frac{11}{12}$ $-\frac{1}{3}$
6. $\frac{2}{7} + \frac{6}{7}$ $1\frac{1}{7}$

7. $\frac{11}{15} + \frac{7}{15}$ $1\frac{1}{5}$
8. $\frac{3}{16} - \frac{1}{16}$ $\frac{1}{8}$
9. $\frac{8}{21} + \frac{5}{21}$ $\frac{13}{21}$

10. $\frac{4}{5} - \frac{3}{4}$ $\frac{1}{20}$
11. $\frac{3}{8} + \frac{1}{2}$ $\frac{7}{8}$
12. $\frac{2}{5} - \frac{21}{25}$ $-\frac{11}{25}$

13. $\frac{11}{12} + \frac{5}{6}$ $1\frac{3}{4}$
14. $\frac{7}{8} - \frac{5}{12}$ $\frac{11}{24}$
15. $\frac{9}{10} + \frac{5}{6}$ $1\frac{11}{15}$

16. $\frac{2}{5} - \frac{7}{8}$ $-\frac{19}{40}$
17. $\frac{5}{6} + \frac{11}{15}$ $1\frac{17}{30}$
18. $\frac{3}{4} - \frac{8}{15}$ $\frac{13}{60}$

19. The school track is $\frac{7}{8}$ mile in length. Sherri ran $\frac{2}{3}$ mile. How much farther does she have to go to get all the way around the track?
$\frac{5}{24}$ mile

20. The Millers budget $\frac{1}{2}$ of their income for fixed expenses and $\frac{1}{8}$ of their income for savings. What fraction of their income is left?
$\frac{3}{8}$

62   **Holt Mathematics**

---

117   **Holt Mathematics**

LESSON 3-8 — Practice C
Adding and Subtracting Fractions

Add or subtract. Write each answer in simplest form.

1. $\frac{7}{15} - \frac{4}{15}$ = $\frac{1}{5}$
2. $\frac{7}{18} + \frac{11}{18}$ = 1
3. $\frac{6}{7} + \frac{8}{21}$ = $1\frac{5}{21}$
4. $\frac{2}{5} + \frac{7}{15}$ = $\frac{13}{15}$
5. $\frac{5}{12} - \frac{4}{9}$ = $-\frac{1}{36}$
6. $\frac{7}{30} - \frac{9}{10}$ = $-\frac{2}{3}$
7. $\frac{8}{9} - \frac{5}{18}$ = $\frac{11}{18}$
8. $\frac{7}{25} + \frac{4}{5}$ = $1\frac{2}{25}$
9. $\frac{3}{8} + \frac{5}{11}$ = $\frac{73}{88}$
10. $\frac{1}{8} - \frac{19}{40}$ = $-\frac{7}{20}$
11. $\frac{5}{8} + \frac{7}{12}$ = $1\frac{5}{24}$
12. $\frac{5}{6} - \frac{5}{9}$ = $\frac{5}{18}$
13. $\frac{7}{8} + \frac{4}{5} + \frac{9}{20}$ = $2\frac{1}{8}$
14. $\frac{11}{12} - \frac{5}{6} - \frac{2}{3}$ = $-\frac{7}{12}$
15. $-\frac{2}{3} + \frac{13}{15} - \frac{4}{5}$ = $-\frac{3}{5}$
16. $-\frac{7}{12} - \frac{1}{5} + \frac{3}{4}$ = $-\frac{1}{30}$
17. $\frac{7}{9} + \frac{2}{5} - \frac{4}{15}$ = $\frac{41}{45}$
18. $\frac{13}{20} - \frac{2}{7} + \frac{11}{35}$ = $\frac{19}{28}$

19. In 2001, the population of the United States was about $\frac{2}{7}$ billion people. It is projected that by 2025, there will be over $\frac{1}{3}$ billion people in the United States. How much will the population increase by 2025? — $\frac{1}{21}$ billion

20. Benjamin walks $\frac{11}{15}$ mile to work and then $\frac{3}{10}$ mile to the grocery store. How far does he walk? — $1\frac{1}{30}$ miles

21. About $\frac{2}{5}$ of the population in the United States has blood type A. About $\frac{23}{50}$ have blood type O. What fraction of the population has either blood type A or O? — about $\frac{43}{50}$

63   **Holt Mathematics**

LESSON 3-8 — Reteach
Adding and Subtracting Fractions

To add or subtract fractions with different denominators:

Step 1: Find the least common multiple of the denominators.

Step 2: Write both fractions with the least common multiple (LCM) as the denominator.

Step 3: Add or subtract the numerators, keeping the denominator the same. Write the answer in simplest form.

$\frac{1}{4} + \frac{5}{6}$ $\frac{2}{3} - \frac{1}{2}$

The LCM of 4 and 6 is 12.   The LCM of 3 and 2 is 6.

$\frac{1}{4} = \frac{3}{12}$ and $\frac{5}{6} = \frac{10}{12}$ $\frac{2}{3} = \frac{4}{6}$ and $\frac{1}{2} = \frac{3}{6}$

$\frac{1}{4} + \frac{5}{6} = \frac{3}{12} + \frac{10}{12} = \frac{3+10}{12} = \frac{13}{12} = 1\frac{1}{12}$ $\frac{2}{3} - \frac{1}{2} = \frac{4}{6} - \frac{3}{6} = \frac{4-3}{6} = \frac{1}{6}$

Add or subtract. Write each answer in simplest form.

1. $\frac{3}{5} + \frac{1}{3} = \frac{9}{15} + \frac{5}{15} = \frac{9+5}{15} = \frac{14}{15}$
2. $\frac{8}{9} - \frac{1}{3} = \frac{8}{9} - \frac{3}{9} = \frac{8-3}{9} = \frac{5}{9}$
3. $\frac{2}{5} + \frac{1}{2} = \frac{4}{10} + \frac{5}{10} = \frac{4+5}{10} = \frac{9}{10}$
4. $\frac{3}{4} - \frac{1}{3} = \frac{9}{12} - \frac{4}{12} = \frac{9-4}{12} = \frac{5}{12}$
5. $\frac{2}{3} + \frac{3}{4} = \frac{8}{12} + \frac{9}{12} = \frac{8+9}{12} = \frac{17}{12} = 1\frac{5}{12}$
6. $\frac{1}{4} - \frac{5}{8} = \frac{2}{8} - \frac{5}{8} = \frac{2-5}{8} = -\frac{3}{8}$
7. $\frac{1}{4} + \frac{2}{3}$ = $\frac{11}{12}$
8. $\frac{9}{10} - \frac{1}{2}$ = $\frac{2}{5}$
9. $\frac{7}{10} + \frac{2}{5}$ = $1\frac{1}{10}$
10. $\frac{11}{12} - \frac{2}{3}$ = $\frac{1}{4}$
11. $\frac{1}{4} - \frac{7}{10}$ = $-\frac{9}{20}$
12. $\frac{3}{4} + \frac{4}{5}$ = $1\frac{11}{20}$

64   **Holt Mathematics**

LESSON 3-8 — Challenge
Fraction Tic-Tac-Toe

Find each sum or difference. Write the answer in simplest terms. Cross off the answers in the tic-tac-toe board to win.

1. $\frac{3}{8} + \frac{1}{2} + \frac{3}{4}$ = $1\frac{5}{8}$
2. $-\frac{6}{7} + \frac{2}{3} + \left(-\frac{11}{21}\right)$ = $-\frac{5}{7}$
3. $\frac{8}{9} + \frac{1}{3} + \left(-\frac{7}{12}\right)$ = $\frac{23}{36}$
4. $\frac{1}{6} + \left(-\frac{2}{3}\right) - \left(-\frac{5}{12}\right)$ = $-\frac{1}{12}$
5. $-\frac{7}{9} - \frac{5}{6} + \frac{1}{3}$ = $-1\frac{5}{18}$
6. $\frac{7}{10} - \left(-\frac{1}{2}\right) + \left(-\frac{3}{5}\right)$ = $\frac{3}{5}$
7. $\frac{9}{14} - \left(-\frac{2}{7}\right) - \left(-\frac{1}{2}\right)$ = $1\frac{3}{7}$
8. $-\frac{8}{15} - \frac{1}{3} + \frac{4}{5}$ = $-\frac{1}{15}$
9. $\frac{2}{9} - \frac{2}{5} + \left(-\frac{2}{15}\right)$ = $-\frac{14}{45}$

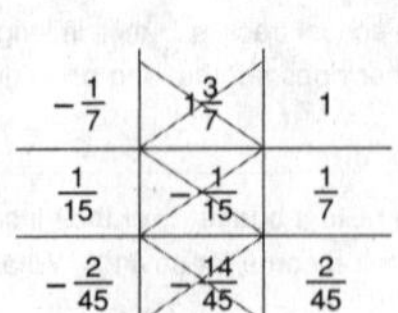

$\frac{1}{2}$	$\frac{1}{7}$	$\frac{23}{36}$
$\frac{2}{12}$	$\frac{5}{7}$	$\frac{5}{7}$
$1\frac{5}{8}$	$-\frac{1}{6}$	$\frac{7}{8}$

$\frac{1}{12}$	1	$-\frac{3}{5}$
$\frac{5}{12}$	$1\frac{5}{18}$	$\frac{4}{5}$
$-\frac{1}{12}$	$-1\frac{5}{18}$	$\frac{3}{5}$

$-\frac{1}{7}$	$1\frac{3}{7}$	1
$\frac{1}{15}$	$\frac{1}{15}$	$\frac{1}{7}$
$-\frac{2}{45}$	$-\frac{14}{45}$	$\frac{2}{45}$

65   **Holt Mathematics**

LESSON 3-8 — Problem Solving
Adding and Subtracting Fractions

Write the correct answer.

1. During the recycling drive, $\frac{1}{5}$ of the material collected was bottles and $\frac{1}{4}$ was paper. Cardboard boxes made up $\frac{1}{10}$ of the material. How much of the total do these three items represent? — $\frac{11}{20}$ of the total

2. Decorations for school dances take $\frac{1}{5}$ of the student council's budget. Entertainment takes $\frac{3}{10}$ of the budget. What fraction of the budget is left? — $\frac{1}{2}$ of the budget

3. The school environmental club made a poster to celebrate Earth Day. The poster is $\frac{7}{8}$ yard long and $\frac{2}{3}$ yard wide. What is the difference in the length and width of the poster? — $\frac{5}{24}$ yard

4. Three students ran for president of the student council. Eddie received $\frac{1}{5}$ of the votes. Tamara received $\frac{3}{8}$ of the votes. Levi received the rest. Which student won the election? — Levi

Choose the letter for the best answer.

5. The Reeds budget $\frac{1}{3}$ of their income for rent and $\frac{1}{4}$ for food. How much of their budget is left?
   A $\frac{1}{2}$
   B $\frac{3}{4}$
   C $\frac{5}{12}$
   D $\frac{7}{12}$

6. Jasmine's CD collection is $\frac{3}{8}$ jazz, $\frac{1}{4}$ classical, and the rest rock music. What fraction of her CDs is rock music?
   F $\frac{1}{4}$
   G $\frac{3}{8}$
   H $\frac{1}{2}$
   J $\frac{5}{8}$

7. Wong has 2 boxes of saltwater taffy. One box contains $\frac{3}{4}$ pound, and the other box contains $\frac{7}{10}$ pound. How much taffy does he have all together?
   A $\frac{9}{10}$ pound
   B $1\frac{9}{20}$ pounds
   C $1\frac{1}{2}$ pounds
   D $1\frac{13}{20}$ pounds

8. In 1992, about $\frac{43}{100}$ people voted for Bill Clinton for President. About $\frac{1}{5}$ voted for Ross Perot and the rest voted for George Bush. About how many voted for George Bush?
   F about $\frac{37}{100}$
   G about $\frac{52}{100}$
   H about $\frac{57}{100}$
   J about $\frac{3}{5}$

66   **Holt Mathematics**

    Holt Mathematics

Reading Strategies
Use Fraction Strips

It is easy to add and subtract fractions with like, or **common** denominators.

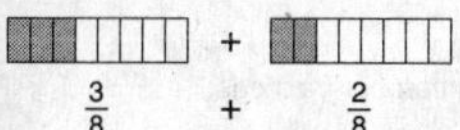

$\frac{3}{8}$ + $\frac{2}{8}$

Use the fraction strips to answer each question.

1. What fractional part of the first fraction strip is shaded? $\frac{3}{8}$

2. What fractional part of the second fraction strip is shaded? $\frac{2}{8}$

3. Add the shaded parts of both fraction strips. What fractional part of both is shaded? $\frac{5}{8}$

4. When you add two fractions with common denominators, does the denominator change? no

$\frac{4}{6}$ − $\frac{3}{6}$

Use the fraction strips to answer each question.

5. What fractional part of the first fraction strip is shaded? $\frac{4}{6}$

6. What fractional part of the second fraction strip is shaded? $\frac{3}{6}$

7. When you subtract the second fraction strip from the first, what is the answer? $\frac{1}{6}$

8. When you subtract two fractions, what part of the fraction is subtracted, the numerator or the denominator? numerator

67

Holt Mathematics

Puzzles, Twisters & Teasers
Jumping to Conclusions!

Solve the equations. Then answer the riddle.

D $\frac{1}{2} - \frac{3}{4}$ $-\frac{1}{4}$

B $\frac{5}{6} - \frac{1}{9}$ $\frac{13}{18}$

E $\frac{1}{2} - \frac{7}{12}$ $-\frac{1}{12}$

C $\frac{4}{5} + \frac{6}{7}$ $1\frac{23}{35}$

O $\frac{5}{7} + \frac{1}{3}$ $1\frac{1}{21}$

A $\frac{3}{4} + \frac{2}{5}$ $1\frac{3}{20}$

L $\frac{1}{2} - \frac{2}{7}$ $\frac{3}{14}$

U $\frac{2}{3} + \frac{1}{3}$ 1

S $\frac{7}{12} + \frac{6}{12}$ $1\frac{1}{12}$

T $\frac{21}{24} - \frac{1}{2}$ $\frac{3}{8}$

I $\frac{1}{5} + \frac{2}{3}$ $\frac{13}{15}$

H $\frac{5}{6} - \frac{1}{6}$ $\frac{2}{3}$

W $\frac{3}{4} - \frac{11}{12}$ $-\frac{1}{6}$

N $\frac{5}{8} + \frac{7}{8}$ $1\frac{1}{2}$

If a cat can jump five feet in the air, then why can't it jump through a two-foot high window?

$\underset{\frac{13}{18}}{B}$ $\underset{-\frac{1}{12}}{E}$ $\underset{1\frac{23}{35}}{C}$ $\underset{1\frac{3}{20}}{A}$ $\underset{1}{U}$ $\underset{1\frac{1}{12}}{S}$ $\underset{-\frac{1}{12}}{E}$

$\underset{\frac{3}{8}}{T}$ $\underset{\frac{2}{3}}{H}$ $\underset{-\frac{1}{12}}{E}$ $\underset{-\frac{1}{6}}{W}$ $\underset{\frac{13}{15}}{I}$ $\underset{1\frac{1}{2}}{N}$ $\underset{-\frac{1}{4}}{D}$ $\underset{1\frac{1}{21}}{O}$ $\underset{-\frac{1}{6}}{W}$

$\underset{\frac{13}{15}}{I}$ $\underset{1\frac{1}{12}}{S}$ $\underset{1\frac{23}{35}}{C}$ $\underset{\frac{3}{14}}{L}$ $\underset{1\frac{1}{21}}{O}$ $\underset{1\frac{1}{12}}{S}$ $\underset{-\frac{1}{12}}{E}$ $\underset{-\frac{1}{4}}{D}$

68

Holt Mathematics

Practice A
Adding and Subtracting Mixed Numbers

Add. Write each answer in simplest form.

1. $5\frac{1}{2} + 2\frac{1}{4}$ $7\frac{3}{4}$

2. $3\frac{1}{4} + 4\frac{3}{4}$ 8

3. $2\frac{1}{5} + 1\frac{2}{5}$ $3\frac{3}{5}$

4. $1\frac{1}{2} + 3\frac{3}{8}$ $4\frac{7}{8}$

5. $2\frac{6}{7} + 5\frac{4}{7}$ $8\frac{3}{7}$

6. $3\frac{2}{9} + 3\frac{8}{9}$ $7\frac{1}{9}$

7. $2\frac{5}{12} + 3\frac{1}{8}$ $5\frac{13}{24}$

8. $2\frac{3}{4} + 5\frac{5}{6}$ $8\frac{7}{12}$

9. $\frac{2}{3} + 2\frac{5}{8}$ $3\frac{7}{24}$

Subtract. Write each answer in simplest form.

10. $4\frac{3}{4} - 2\frac{1}{4}$ $2\frac{1}{2}$

11. $5\frac{5}{6} - 3\frac{1}{6}$ $2\frac{2}{3}$

12. $7\frac{2}{3} - 4\frac{1}{3}$ $3\frac{1}{3}$

13. $5\frac{1}{4} - 1\frac{3}{4}$ $3\frac{1}{2}$

14. $6\frac{1}{3} - 5\frac{2}{3}$ $\frac{2}{3}$

15. $8\frac{1}{6} - 5\frac{5}{6}$ $2\frac{1}{3}$

16. $9\frac{5}{6} - 6\frac{1}{4}$ $3\frac{7}{12}$

17. $7\frac{3}{4} - 4\frac{5}{6}$ $2\frac{11}{12}$

18. $8\frac{3}{8} - 4\frac{3}{4}$ $3\frac{5}{8}$

19. Samson bicycled $8\frac{7}{8}$ miles on Friday and $5\frac{1}{4}$ miles on Saturday. How much farther did he ride on Friday?

$3\frac{5}{8}$ miles

69

Holt Mathematics

Practice B
Adding and Subtracting Mixed Numbers

Add. Write each answer in simplest form.

1. $7\frac{2}{7} + 6\frac{5}{7}$ 14

2. $5\frac{4}{9} + 3\frac{7}{9}$ $9\frac{2}{9}$

3. $4\frac{1}{3} + 8\frac{1}{4}$ $12\frac{7}{12}$

4. $2\frac{7}{15} + 3\frac{11}{15}$ $6\frac{1}{5}$

5. $6\frac{9}{10} + 1\frac{2}{5}$ $8\frac{3}{10}$

6. $2\frac{3}{5} + 1\frac{11}{20}$ $4\frac{3}{20}$

7. $5\frac{9}{10} + 2\frac{5}{8}$ $8\frac{21}{40}$

8. $2\frac{11}{12} + 3\frac{7}{8}$ $6\frac{19}{24}$

9. $1\frac{2}{3} + 5\frac{7}{9}$ $7\frac{4}{9}$

Subtract. Write each answer in simplest form.

10. $7\frac{7}{9} - 3\frac{5}{9}$ $4\frac{2}{9}$

11. $9\frac{7}{10} - 5\frac{3}{10}$ $4\frac{2}{5}$

12. $4\frac{13}{15} - 1\frac{7}{15}$ $3\frac{2}{5}$

13. $6\frac{2}{3} - 3\frac{3}{5}$ $3\frac{1}{15}$

14. $10\frac{3}{4} - 6\frac{1}{3}$ $4\frac{5}{12}$

15. $2\frac{3}{10} - 1\frac{7}{8}$ $\frac{17}{40}$

16. $8\frac{7}{12} - 6\frac{1}{3}$ $2\frac{1}{4}$

17. $5\frac{7}{8} - 3\frac{9}{10}$ $1\frac{39}{40}$

18. $7\frac{6}{7} - 6\frac{3}{4}$ $1\frac{3}{28}$

19. Tucker ran $5\frac{3}{8}$ miles on Monday and $3\frac{3}{4}$ miles on Tuesday. How far did he run on both days?

$9\frac{1}{8}$ miles

70

Holt Mathematics

Practice C

LESSON 3-9

Adding and Subtracting Mixed Numbers

Add or subtract. Write each answer in simplest form.

1. $3\frac{11}{21} + 2\frac{8}{21}$
$5\frac{19}{21}$

2. $2\frac{5}{26} + 5\frac{11}{13}$
$8\frac{1}{26}$

3. $4\frac{2}{11} + 1\frac{6}{22}$
$5\frac{5}{11}$

4. $15\frac{16}{19} - 8\frac{7}{19}$
$7\frac{9}{19}$

5. $9\frac{13}{25} - 7\frac{2}{5}$
$2\frac{3}{25}$

6. $4\frac{11}{20} - 3\frac{3}{10}$
$1\frac{1}{4}$

7. $7\frac{3}{5} + 8\frac{1}{8}$
$15\frac{29}{40}$

8. $4\frac{2}{3} + 6\frac{3}{7}$
$11\frac{2}{21}$

9. $9\frac{5}{12} + 3\frac{3}{8}$
$12\frac{19}{24}$

10. $8\frac{5}{6} - 5\frac{3}{5}$
$3\frac{7}{30}$

11. $7\frac{7}{12} - 4\frac{11}{15}$
$2\frac{17}{20}$

12. $10\frac{2}{3} - 5\frac{13}{18}$
$4\frac{17}{18}$

13. $3\frac{2}{5} + 4\frac{1}{3} - 6\frac{7}{10}$
$1\frac{1}{30}$

14. $1\frac{1}{4} + 6\frac{7}{8} + 5\frac{5}{16}$
$13\frac{7}{16}$

15. $-4\frac{1}{2} - 9\frac{5}{6} - 5\frac{3}{4}$
$-20\frac{1}{12}$

16. $2\frac{1}{12} - 3\frac{2}{5} + 7\frac{4}{15}$
$5\frac{19}{20}$

17. $1\frac{7}{8} + 2\frac{7}{20} - 1\frac{3}{10}$
$2\frac{37}{40}$

18. $-6\frac{6}{7} - 3\frac{8}{35} - 1\frac{2}{5}$
$-11\frac{17}{35}$

Compare. Write <, >, or =.

19. $11\frac{1}{5} - 5\frac{4}{5} \boxed{<} 7\frac{1}{3} - 1\frac{1}{2}$

20. $7\frac{2}{3} + 4\frac{1}{5} \boxed{<} 7\frac{5}{6} + 4\frac{2}{7}$

21. $18\frac{1}{2} - 3\frac{3}{4} \boxed{=} 7\frac{2}{3} + 7\frac{1}{12}$

22. $6\frac{1}{2} - 3\frac{1}{3} \boxed{>} 5\frac{1}{3} - 2\frac{1}{5}$

23. The men's indoor pole vault record was set in 1993 at $20\frac{1}{6}$ feet. The women's record was set in 2001 at $15\frac{5}{12}$ feet. How much higher is the men's record than the women's record?

$4\frac{3}{4}$ feet

24. Richard set a goal of running 10 miles per week. On Monday he ran $3\frac{3}{5}$ miles. On Friday he ran $3\frac{9}{10}$ miles. How much farther does he still have to run to meet his weekly goal?

$2\frac{1}{2}$ miles

71

Holt Mathematics

Reteach

LESSON 3-9

Adding and Subtracting Mixed Numbers

You can write mixed numbers as improper fractions before adding.

Add: $4\frac{5}{8} + 2\frac{7}{8}$

- Write improper fractions.

  $4\frac{5}{8} = \frac{37}{8}$ and $2\frac{7}{8} = \frac{23}{8}$

- The denominators are the same. Add and simplify.

  $\frac{37}{8} + \frac{23}{8} = \frac{60}{8} = 7\frac{4}{8} = 7\frac{1}{2}$

Add: $1\frac{2}{5} + 3\frac{3}{4}$

- Write improper fractions.

  $1\frac{2}{5} = \frac{7}{5}$ and $3\frac{3}{4} = \frac{15}{4}$

- Find a common denominator. The LCD is 20.

  $\frac{7}{5} = \frac{28}{20}$ and $\frac{15}{4} = \frac{75}{20}$

- Add and simplify.

  $\frac{28}{20} + \frac{75}{20} = \frac{103}{20} = 5\frac{3}{20}$

Add. Write each answer in simplest form.

1. $4\frac{7}{10} + 2\frac{9}{10} = \frac{47}{10} + \frac{29}{10} = \frac{76}{10} = 7\frac{6}{10} = 7\frac{3}{5}$

2. $2\frac{1}{2} + 1\frac{3}{8} = \frac{5}{2} + \frac{11}{8} = \frac{20}{8} + \frac{11}{8} = \frac{31}{8} = 3\frac{7}{8}$

3. $3\frac{1}{5} + 2\frac{1}{3} = \frac{16}{5} + \frac{7}{3} = \frac{48}{15} + \frac{35}{15} = \frac{83}{15} = 5\frac{8}{15}$

4. $1\frac{2}{7} + 5\frac{3}{7}$
$6\frac{5}{7}$

5. $5\frac{3}{8} + 3\frac{7}{8}$
$9\frac{1}{4}$

6. $4\frac{4}{9} + 2\frac{2}{3}$
$7\frac{1}{9}$

7. $2\frac{3}{5} + 3\frac{7}{10}$
$6\frac{3}{10}$

8. $2\frac{3}{4} + 1\frac{5}{6}$
$4\frac{7}{12}$

9. $4\frac{1}{3} + 2\frac{1}{2}$
$6\frac{5}{6}$

72

Holt Mathematics

Reteach

LESSON 3-9

Adding and Subtracting Mixed Numbers (continued)

You can write whole numbers as fractions and mixed numbers as improper fractions before subtracting.

$6 - 2\frac{5}{9}$

- Write 6 as a fraction and $2\frac{5}{9}$ as an improper fraction.

  $6 = \frac{6}{1}$ and $2\frac{5}{9} = \frac{23}{9}$

- Find a common denominator. The LCD is 9.

  $\frac{6}{1} = \frac{54}{9}$

- Subtract and simplify.

  $\frac{54}{9} - \frac{23}{9} = \frac{31}{9} = 3\frac{4}{9}$

$4\frac{1}{4} - 1\frac{5}{6}$

- Write improper fractions.

  $4\frac{1}{4} = \frac{17}{4}$ and $1\frac{5}{6} = \frac{11}{6}$

- Find a common denominator. The LCD is 12.

  $\frac{17}{4} = \frac{51}{12}$ and $\frac{11}{6} = \frac{22}{12}$

- Subtract and simplify.

  $\frac{51}{12} - \frac{22}{12} = \frac{29}{12} = 2\frac{5}{12}$

Subtract. Write each answer in simplest form.

10. $6\frac{1}{4} - 1\frac{3}{4} = \frac{25}{4} - \frac{7}{4} = \frac{18}{4} = 4\frac{2}{4} = 4\frac{1}{2}$

11. $5 - 1\frac{2}{3} = \frac{5}{1} - \frac{5}{3} = \frac{15}{3} - \frac{5}{3} = \frac{10}{3} = 3\frac{1}{3}$

12. $5\frac{3}{4} - 4\frac{1}{8} = \frac{23}{4} - \frac{33}{8} = \frac{46}{8} - \frac{33}{8} = \frac{13}{8} = 1\frac{5}{8}$

13. $3\frac{1}{4} - 2\frac{5}{6} = \frac{13}{4} - \frac{17}{6} = \frac{39}{12} - \frac{34}{12} = \frac{5}{12}$

14. $8\frac{2}{5} - 6\frac{4}{5}$
$1\frac{3}{5}$

15. $5 - 3\frac{4}{7}$
$1\frac{3}{7}$

16. $10 - 5\frac{7}{10}$
$4\frac{3}{10}$

17. $7\frac{1}{2} - 6\frac{3}{10}$
$1\frac{1}{5}$

18. $2\frac{2}{3} - 1\frac{2}{5}$
$1\frac{4}{15}$

19. $5\frac{1}{2} - 3\frac{2}{3}$
$1\frac{5}{6}$

73

Holt Mathematics

Challenge

LESSON 3-9

Mixed Number Magic

In a magic square, each row, column, and diagonal has the same sum. Fill in the missing numbers to complete each magic square below.

Magic Square #1

$4\frac{1}{3}$	$9\frac{3}{4}$	$2\frac{1}{6}$
$3\frac{1}{4}$	$5\frac{5}{12}$	$7\frac{7}{12}$
$8\frac{2}{3}$	$1\frac{1}{12}$	$6\frac{1}{2}$

Magic Square #1 sum: $16\frac{1}{4}$

Magic Square #2

$2\frac{1}{6}$	$4\frac{7}{8}$	$1\frac{1}{12}$
$1\frac{5}{8}$	$2\frac{17}{24}$	$3\frac{19}{24}$
$4\frac{1}{3}$	$\frac{13}{24}$	$3\frac{1}{4}$

Magic Square #2 sum: $8\frac{1}{8}$

What is the difference between these two magic sums? $8\frac{1}{8}$

Magic Square #3

6	$13\frac{1}{2}$	3
$4\frac{1}{2}$	$7\frac{1}{2}$	$10\frac{1}{2}$
12	$1\frac{1}{2}$	9

Magic Square #3 sum: $22\frac{1}{2}$

Magic Square #4

$1\frac{1}{2}$	$3\frac{3}{8}$	$\frac{3}{4}$
$1\frac{1}{8}$	$1\frac{7}{8}$	$2\frac{5}{8}$
3	$\frac{3}{8}$	$2\frac{1}{4}$

Magic Square #4 sum: $5\frac{5}{8}$

What is the difference between these two magic sums? $16\frac{7}{8}$

74

Holt Mathematics

Holt Mathematics

Problem Solving
Adding and Subtracting Mixed Numbers

Write the correct answer.

1. A female gray whale is $45\frac{1}{4}$ feet long. A male gray whale is $43\frac{1}{2}$ feet long. How much longer is the female than the male gray whale?

   $1\frac{3}{4}$ feet

2. At birth, a pilot whale is $4\frac{3}{5}$ feet long. A newborn gray whale is $15\frac{1}{4}$ feet long. How much longer is the newborn gray whale?

   $10\frac{13}{20}$ feet

3. A manatee weighs $\frac{1}{2}$ ton. A walrus weighs $1\frac{3}{4}$ tons. A narwhal weighs $1\frac{1}{2}$ tons. What is the total weight of all 3 animals?

   $3\frac{3}{4}$ tons

4. A bottle-nosed dolphin can leap $15\frac{1}{8}$ feet out of the water. The world record high jump for a human is $8\frac{1}{24}$ feet. How much higher can a dolphin leap than a human?

   $7\frac{1}{12}$ feet

Choose the letter for the best answer.

5. At a wildlife park, the killer whale show lasts $\frac{5}{8}$ of an hour. The guided tour of the park takes $2\frac{1}{2}$ hours. How long will it take to do both activities?

   (A) $3\frac{1}{8}$ hours
   B $2\frac{5}{8}$ hours
   C $3\frac{5}{8}$ hours
   D $1\frac{7}{8}$ hours

6. Jeremy walks $3\frac{1}{2}$ miles while visiting a wildlife park. Shawna hikes a $4\frac{3}{8}$-mile long nature trail. How much farther does Shawna walk?

   F $\frac{3}{8}$ of a mile
   G $\frac{5}{8}$ of a mile
   (H) $\frac{7}{8}$ of a mile
   J $1\frac{1}{8}$ of a mile

7. Dog food comes in $5\frac{7}{8}$-pound bags and $12\frac{3}{4}$-pound bags. Find the total weight of 2 small and 1 large bags.

   A $22\frac{7}{8}$ pounds
   B $23\frac{1}{2}$ pounds
   (C) $24\frac{1}{2}$ pounds
   D $25\frac{3}{4}$ pounds

8. A movie lasts $2\frac{1}{6}$ hours. A baseball game lasts $3\frac{2}{3}$ hours. How much longer does the game last?

   F $\frac{5}{6}$ hours
   G $1\frac{1}{6}$ hours
   H $1\frac{1}{3}$ hours
   (J) $1\frac{1}{2}$ hours

Reading Strategies
Follow a Procedure

To add mixed numbers that have fractions with common denominators, follow these steps.

$$2\frac{3}{7} + 1\frac{5}{7}$$

Step 1: Add fractions.
$$\begin{array}{r} 2\frac{3}{7} \\ +\ 1\frac{5}{7} \\ \hline \frac{8}{7} \end{array}$$

Step 2: Add integers.
$$\begin{array}{r} 2\frac{3}{7} \\ +\ 1\frac{5}{7} \\ \hline 3 \end{array}$$

Step 3: Simplify.
$$\begin{array}{r} 2\frac{3}{7} \\ +\ 1\frac{5}{7} \\ \hline 3\frac{8}{7} = 4\frac{1}{7} \end{array}$$

Use the problem above to answer each question.

1. What part of the mixed numbers should you add first?   **fractions**

2. What part of the mixed numbers should you add next?   **integers**

3. Do you need to simplify the sum?   **yes**

When you subtract mixed numbers, you use the same steps as when you add them. However, when the step says add fractions or integers, you will subtract.

Use the problem $5\frac{3}{4} - 2\frac{1}{4}$ and the steps above to answer each question.

4. What is the first step in solving the problem?

   subtract fractions

5. What is the second step in solving the problem?

   subtract integers

6. What is the answer of the problem?

   $3\frac{2}{4}$, or $3\frac{1}{2}$

Puzzles, Twisters & Teasers
Reaching the Right Solutions!

Decide whether or not each equation is correct. Circle the letters above your answers. Then answer the riddle.

1. $1\frac{2}{15} + 7\frac{1}{6} = 8\frac{3}{10}$
   (T) correct   B incorrect

2. $1\frac{7}{9} - \frac{17}{18} = 1$
   Z correct   (O) incorrect

3. $70\frac{4}{5} + 1\frac{4}{5} = 72\frac{3}{5}$
   (R) correct   A incorrect

4. $7\frac{1}{3} + 8\frac{1}{5} = 16\frac{2}{5}$
   Y correct   (E) incorrect

5. $6\frac{1}{4} + 8\frac{3}{4} = 14$
   X correct   (A) incorrect

6. $3\frac{4}{5} + 3\frac{2}{5} = 7\frac{1}{5}$
   (C) correct   P incorrect

7. $3\frac{1}{2} + 5\frac{1}{4} = 8\frac{3}{4}$
   (H) correct   V incorrect

8. $14\frac{3}{5} - 8\frac{1}{2} = 7\frac{1}{5}$
   U correct   (H) incorrect

9. $6\frac{3}{4} - 2\frac{3}{4} = 4$
   (I) correct   Q incorrect

10. $9\frac{1}{6} - 4\frac{6}{9} = 5\frac{1}{3}$
    M correct   (S) incorrect

11. $6\frac{1}{6} + 5\frac{3}{10} = 12\frac{1}{2}$
    L correct   (H) incorrect

12. $8 - 2\frac{3}{4} = 5\frac{1}{4}$
    (A) correct   K incorrect

13. $3\frac{1}{3} - 2\frac{5}{8} = 6\frac{7}{8}$
    J correct   (N) incorrect

14. $1\frac{8}{9} + 4\frac{4}{9} = 5\frac{9}{9}$
    G correct   (D) incorrect

15. $6\frac{2}{3} - 5\frac{1}{3} = 1\frac{1}{3}$
    (S) correct   F incorrect

Why did the basketball player need long arms?

T O   R E A C H
H I S   H A N D S .

Practice A
Multiplying Fractions and Mixed Numbers

Multiply. Choose the letter for the best answer.

1. $7 \cdot \frac{1}{8}$
   A $\frac{1}{56}$ C $\frac{3}{4}$
   B $\frac{7}{15}$ (D) $\frac{7}{8}$

2. $\frac{2}{5} \cdot \frac{3}{4}$
   F $\frac{1}{4}$ (H) $\frac{3}{10}$
   G $\frac{2}{3}$ J $\frac{5}{9}$

3. $4 \cdot 3\frac{3}{5}$
   A $2\frac{2}{5}$ C $13\frac{1}{5}$
   B 12 (D) $14\frac{2}{5}$

4. $1\frac{1}{4} \cdot 2\frac{2}{3}$
   F $2\frac{1}{6}$ H $3\frac{2}{3}$
   (G) $3\frac{1}{3}$ J $3\frac{11}{12}$

Multiply. Write each answer in simplest form.

5. $4 \cdot \frac{1}{2}$

   2

6. $8 \cdot \frac{1}{4}$

   2

7. $10 \cdot \frac{1}{5}$

   2

8. $\frac{1}{2} \cdot \frac{1}{4}$

   $\frac{1}{8}$

9. $\frac{1}{4} \cdot -\frac{2}{3}$

   $\left(-\frac{1}{6}\right)$

10. $\frac{3}{4} \cdot \frac{2}{3}$

    $\frac{1}{2}$

11. $-16 \cdot \frac{3}{4}$

    -12

12. $24 \cdot \frac{5}{6}$

    20

13. $32 \cdot \frac{3}{8}$

    12

14. $2\frac{1}{4} \cdot \frac{1}{2}$

    $1\frac{1}{8}$

15. $3\frac{1}{3} \cdot \frac{3}{5}$

    2

16. $5\frac{1}{3} \cdot \frac{1}{4}$

    $1\frac{1}{3}$

17. $1\frac{1}{2} \cdot 1\frac{1}{5}$

    $1\frac{4}{5}$

18. $1\frac{2}{5} \cdot 2\frac{3}{4}$

    $3\frac{17}{20}$

19. $2\frac{2}{7} \cdot 3\frac{1}{8}$

    $7\frac{1}{7}$

20. Louis spent 12 hours last week practicing guitar. If $\frac{1}{4}$ of the time was spent practicing chords, how much time did he spend practicing chords?

    3 hours

21. A banana bread recipe calls for $\frac{1}{4}$ tsp salt. Diane is making 5 loaves of banana bread. How much salt does she need?

    $1\frac{1}{4}$ tsp

Practice B
Multiplying Fractions and Mixed Numbers

Multiply. Write each answer in simplest form.

1. $5 \cdot \frac{1}{2}$

$2\frac{1}{2}$

2. $9 \cdot \frac{3}{4}$

$6\frac{3}{4}$

3. $6 \cdot -\frac{2}{5}$

$-2\frac{2}{5}$

4. $\frac{9}{15} \cdot \frac{5}{7}$

$\frac{3}{7}$

5. $\frac{9}{14} \cdot -\frac{7}{9}$

$-\frac{1}{2}$

6. $\frac{7}{12} \cdot \frac{6}{14}$

$\frac{1}{4}$

7. $-12 \cdot \frac{3}{7}$

$-5\frac{1}{7}$

8. $15 \cdot \frac{5}{6}$

$12\frac{1}{2}$

9. $21 \cdot \frac{3}{8}$

$7\frac{7}{8}$

10. $2\frac{1}{3} \cdot \frac{3}{5}$

$1\frac{2}{5}$

11. $3\frac{2}{5} \cdot \frac{1}{2}$

$1\frac{7}{10}$

12. $4\frac{5}{6} \cdot \frac{2}{5}$

$1\frac{14}{15}$

13. $2\frac{2}{5} \cdot \frac{2}{3}$

$1\frac{3}{5}$

14. $3\frac{3}{4} \cdot \frac{2}{5}$

$1\frac{1}{2}$

15. $8\frac{1}{6} \cdot \frac{3}{7}$

$3\frac{1}{2}$

16. $2\frac{1}{3} \cdot 3\frac{3}{8}$

$7\frac{7}{8}$

17. $1\frac{3}{5} \cdot 6\frac{2}{3}$

$10\frac{2}{3}$

18. $2\frac{2}{5} \cdot 4\frac{5}{6}$

$11\frac{3}{5}$

19. Rolf spent 15 hours last week practicing his saxophone. If $\frac{3}{10}$ of the time was spent practicing warm-up routines, how much time did he spend practicing warm-up routines?

$4\frac{1}{2}$ hours

20. A muffin recipe calls for $\frac{2}{5}$ tablespoon of vanilla extract for 6 muffins. Arthur is making 18 muffins. How much vanilla extract does he need?

$1\frac{1}{5}$ T

79
Holt Mathematics

Practice C
Multiplying Fractions and Mixed Numbers

Multiply. Write each answer in simplest form.

1. $12 \cdot \frac{1}{7}$

$1\frac{5}{7}$

2. $15 \cdot \frac{1}{4}$

$3\frac{3}{4}$

3. $-7 \cdot \frac{1}{5}$

$-1\frac{2}{5}$

4. $\frac{8}{15} \cdot \left(-\frac{3}{4}\right)$

$-\frac{2}{5}$

5. $\frac{6}{15} \cdot \frac{7}{18}$

$\frac{7}{45}$

6. $\frac{9}{11} \cdot \frac{22}{27}$

$\frac{2}{3}$

7. $\frac{2}{9} \cdot \frac{7}{10}$

$\frac{7}{45}$

8. $\frac{2}{15} \cdot \frac{5}{12}$

$\frac{1}{18}$

9. $\frac{7}{12} \cdot \frac{3}{7} \cdot \frac{1}{2}$

$\frac{1}{8}$

10. $3\frac{1}{2} \cdot 2\frac{5}{7}$

$9\frac{1}{2}$

11. $3\frac{3}{8} \cdot 4\frac{2}{5}$

$14\frac{17}{20}$

12. $-4\frac{1}{8} \cdot \frac{2}{9}$

$-\frac{11}{12}$

13. $2\frac{2}{11} \cdot \frac{2}{3}$

$1\frac{5}{11}$

14. $3\frac{3}{5} \cdot \frac{2}{9} \cdot \frac{2}{3}$

$\frac{8}{15}$

15. $5\frac{2}{5} \cdot 3\frac{1}{9}$

$16\frac{4}{5}$

Complete each multiplication sentence.

16. $\frac{?}{5} \cdot \frac{3}{10} = \frac{3}{25}$

2

17. $\frac{1}{4} \cdot \frac{?}{7} = \frac{3}{14}$

6

18. $\frac{2}{3} \cdot \frac{?}{8} = \frac{5}{12}$

5

19. $\frac{?}{3} \cdot \frac{3}{8} = \frac{1}{4}$

2

20. $\frac{4}{5} \cdot \frac{?}{4} = \frac{3}{5}$

3

21. $\frac{3}{4} \cdot \frac{?}{7} = \frac{3}{7}$

4

22. A hippopotamus lives about $2\frac{4}{7}$ times as long as a tiger. A tiger lives an average of 16 years. How long does the average hippopotamus live?

$41\frac{1}{7}$ years

23. A pie crust recipe calls for $\frac{4}{5}$ teaspoon of baking powder. Alice is making 13 pies. How much baking powder does she need?

$10\frac{2}{5}$ teaspoons

80
Holt Mathematics

Reteach
Multiplying Fractions and Mixed Numbers

To multiply fractions and mixed numbers:
Step 1: Write any mixed numbers as improper fractions.
Step 2: Multiply the numerators.
Step 3: Multiply the denominators.
Step 4: Write the answer in simplest form.

Remember, positive times negative equals negative.

Multiply: $\frac{4}{9} \cdot \frac{3}{8}$

$\frac{4}{9} \cdot \frac{3}{8} = \frac{4 \cdot 3}{9 \cdot 8}$

Divide numerator and denominator by 12, the GCF.

$= \frac{12}{72}$

$= \frac{1}{6}$

Multiply: $6\frac{1}{4} \cdot \left(-1\frac{4}{5}\right)$

$6\frac{1}{4} \cdot \left(-1\frac{4}{5}\right) = \frac{25}{4} \cdot \left(\frac{-9}{5}\right)$

$= \frac{25 \cdot (-9)}{4 \cdot 5}$

$= \frac{-225}{20}$

$= -11\frac{1}{4}$

Multiply. Write each answer in simplest form.

1. $6 \cdot \frac{1}{9} = \frac{6 \cdot 1}{9} = \frac{6}{9} = \frac{2}{3}$

2. $-\frac{4}{5} \cdot \frac{5}{7} = -\frac{4 \cdot 5}{5 \cdot 7} = -\frac{20}{35} = -\frac{4}{7}$

3. $3\frac{1}{3} \cdot 9 = \frac{10}{3} \cdot 9 = \frac{10 \cdot 9}{3} = \frac{90}{3} = 30$

4. $\frac{3}{10} \cdot 2\frac{1}{2} = \frac{3}{10} \cdot \frac{5}{2} = \frac{3 \cdot 5}{10 \cdot 2} = \frac{15}{20} = \frac{3}{4}$

5. $\frac{2}{7} \cdot \frac{7}{8}$

$\frac{1}{4}$

6. $-\frac{5}{9} \cdot \frac{3}{4}$

$-\frac{5}{12}$

7. $\frac{9}{10} \cdot \left(-\frac{2}{3}\right)$

$-\frac{3}{5}$

8. $2\frac{5}{8} \cdot \frac{2}{3}$

$1\frac{3}{4}$

9. $\frac{1}{2} \cdot 4\frac{1}{4}$

$2\frac{1}{8}$

10. $-\frac{2}{3} \cdot 1\frac{3}{4}$

$-1\frac{1}{6}$

11. $5\frac{1}{5} \cdot \left(-1\frac{2}{3}\right)$

$-8\frac{2}{3}$

12. $4\frac{1}{2} \cdot 1\frac{1}{9}$

5

13. $-2\frac{3}{4} \cdot \left(-1\frac{1}{3}\right)$

$3\frac{2}{3}$

81
Holt Mathematics

Challenge
Weigh Out in Space

How much do you think you would weigh on Mars? You can calculate anyone's weight on any of the planets by multiplying his or her weight times the planet's gravitational pull. Use the table to calculate what each animal listed below would weigh on different planets. Write your answers in simplest form.

Gravitational Pull of the Planets

Planet	Mercury	Venus	Mars	Jupiter	Saturn	Uranus	Neptune
Gravitational Pull	$\frac{19}{50}$	$\frac{9}{10}$	$\frac{19}{50}$	$2\frac{1}{2}$	$\frac{19}{20}$	$\frac{4}{5}$	$1\frac{1}{5}$

	Animal	Weight on Earth	Planet	Weight on Other Planet
1.	Mouse	$\frac{1}{16}$ pound	Neptune	$\frac{3}{40}$ pounds
2.	Cat	$5\frac{1}{4}$ pounds	Mercury	$1\frac{199}{200}$ pounds
			Jupiter	$13\frac{1}{8}$ pounds
3.	Dog	$50\frac{1}{2}$ pounds	Saturn	$47\frac{39}{40}$ pounds
			Neptune	$60\frac{3}{5}$ pounds
4.	Gorilla	200 pounds	Mars	76 pounds
			Uranus	160 pounds
5.	Cow	1,000 pounds	Venus	900 pounds
			Jupiter	2,500 pounds
6.	Small parrot	$\frac{7}{8}$ pound	Mercury	$\frac{133}{400}$ pounds
			Neptune	$1\frac{1}{20}$ pounds
7.	Macaw	$3\frac{1}{2}$ pounds	Neptune	$4\frac{1}{5}$ pounds
8.	Elephant	10,700 pounds	Uranus	8,560 pounds

82
Holt Mathematics

122
Holt Mathematics

Problem Solving
3-10 Multiplying Fractions and Mixed Numbers

Write the correct answer.

1. Ariel's English homework is to read 24 pages. She reads $\frac{1}{8}$ of the assignment on the bus ride home. How many pages does she read on the bus?

   3 pages

2. When a group of 40 friends goes to the movies at a multiplex, $\frac{1}{5}$ of the group decides to watch a science fiction movie. How many of the group see the science fiction movie?

   8 friends

3. As of 1990, the American Indian population was about 2,000,000. About $\frac{1}{5}$ were Cherokee. About how many members of the Cherokee tribe were there in 1990?

   400,000 members

4. Ron spends 3 hours painting a picture. Ashley spends $2\frac{2}{3}$ as long creating a sculpture. How long does Ashley work on her sculpture?

   8 hours

Choose the letter for the best answer.

5. One cup of dry dog food weighs $1\frac{4}{5}$ ounces. A K9 dog eats $6\frac{1}{3}$ cups of food a day. How many ounces of food does the dog eat each day?

   A $4\frac{8}{15}$ ounces Ⓒ $11\frac{2}{5}$ ounces

   B $8\frac{2}{15}$ ounces D $3\frac{14}{27}$ ounces

6. Max has enough chicken to make $8\frac{1}{2}$ servings of salad. He needs $\frac{1}{8}$ pound of chicken per serving. How much chicken does he have?

   F 8 pounds H $8\frac{5}{8}$ pounds

   Ⓖ $1\frac{1}{16}$ pounds J $2\frac{1}{2}$ pounds

7. A meteorite found in Willamette, Oregon, weighed $\frac{7}{10}$ as much as one found in Armanti, Western Mongolia. The meteorite found in Armanti weighed 22 tons. How much did the one in Oregon weigh?

   A $31\frac{1}{5}$ tons

   B $21\frac{3}{10}$ tons

   C $22\frac{7}{10}$ tons

   Ⓓ $15\frac{2}{5}$ tons

8. Nicole takes part in a $12\frac{1}{2}$-mile walk-a-thon to raise money for charity. She stops $\frac{1}{2}$ of the way to rest. How much farther must Nicole walk to finish the walk-a-thon?

   Ⓕ $6\frac{1}{4}$ miles

   G $6\frac{1}{2}$ miles

   H 12 miles

   J 25 miles

Holt Mathematics

Reading Strategies
3-10 Use Fraction Strips

You can write a multiplication problem as a repeated addition problem.

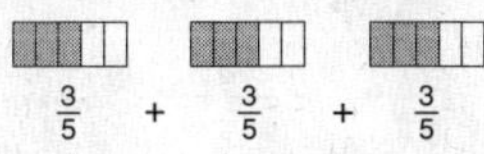

$\frac{3}{5}$ + $\frac{3}{5}$ + $\frac{3}{5}$

$\frac{3}{5} + \frac{3}{5} + \frac{3}{5}$ ◄— Repeated addition

three times three-fifths ◄— Multiplication

$3 \cdot \frac{3}{5}$

1. What fractional part of the fraction strips is shaded? _____ $\frac{3}{5}$

2. How many fraction strips are there? _____ 3

3. Count the number of fractional parts that are shaded in all. How many are there? _____ $\frac{9}{5}$

4. How can you find the answer to the problem above using addition?

   Add the numerators.

You can also find the answer to the above problem using multiplication.

$3 \times \frac{3}{5} = \frac{9}{5}$

5. What fractional part of each fraction strip is shaded? _____ $\frac{2}{3}$

6. How many of these fraction strips are there? _____ 4

7. Write a multiplication equation for this picture. _____ $\frac{2}{3} \times 4 = \frac{8}{3}$

Holt Mathematics

Puzzles, Twisters & Teasers
3-10 Wooden It Be Nice?

Decide whether or not each equation is correct. Circle the letter above your answer. Use the letters you circled to solve the riddle.

1. $2 \cdot \frac{3}{8} = 0.75$

   Ⓘ correct   L incorrect

6. $\frac{1}{2} \cdot \frac{1}{3} \cdot \frac{1}{4} = \frac{1}{16}$

   R correct   Ⓓ incorrect

2. $-15 \cdot \frac{2}{3} = -12$

   M correct   Ⓣ incorrect

7. $\frac{1}{6} \cdot 12\frac{1}{2} = 24$

   S correct   Ⓔ incorrect

3. $\frac{1}{4} \cdot \frac{4}{5} = \frac{1}{5}$

   Ⓦ correct   N incorrect

8. $2\frac{1}{4} \cdot 1\frac{1}{2} = 3\frac{3}{8}$

   Ⓝ correct   J incorrect

4. $\frac{1}{3} \cdot 4\frac{1}{2} = 1\frac{1}{2}$

   Ⓞ correct   P incorrect

9. $2 \cdot 5\frac{2}{3} = 10\frac{1}{3}$

   H correct   Ⓖ incorrect

5. $3\frac{3}{5} \cdot 1\frac{1}{12} = 3\frac{7}{10}$

   Q correct   Ⓞ incorrect

10. $75 \cdot 1\frac{1}{4} = 93.75$

   Ⓞ correct   K incorrect

What happened to the wooden plane with the wooden wheels and wooden engine?

I T

W O O D E N

G O

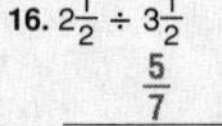

Holt Mathematics

Practice A
3-11 Dividing Fractions and Mixed Numbers

Divide. Write each answer in simplest form.

1. $5 \div \frac{1}{2}$ _____ **10**

2. $9 \div \frac{1}{3}$ _____ **27**

3. $6 \div \frac{1}{4}$ _____ **24**

4. $3 \div \frac{3}{4}$ _____ **4**

5. $10 \div \frac{5}{6}$ _____ **12**

6. $6 \div \frac{3}{8}$ _____ **16**

Divide. Find each quotient in the box.

| $\frac{1}{5}$ | $\left(-\frac{1}{4}\right)$ | $\frac{1}{2}$ | $\left(-\frac{6}{11}\right)$ | $\frac{5}{7}$ | $\frac{7}{8}$ | 1 | $1\frac{1}{2}$ | 2 | $2\frac{6}{7}$ | 3 | 4 | $5\frac{1}{3}$ | $7\frac{1}{2}$ |

7. $\frac{9}{5} \div \frac{3}{5}$ _____ **3**

8. $\frac{6}{7} \div \frac{3}{7}$ _____ **2**

9. $\frac{1}{6} \div \frac{5}{6}$ _____ $\mathbf{\frac{1}{5}}$

10. $\frac{1}{3} \div \frac{2}{3}$ _____ $\mathbf{\frac{1}{2}}$

11. $\frac{3}{4} \div \frac{1}{2}$ _____ $\mathbf{1\frac{1}{2}}$

12. $\frac{1}{6} \div \left(-\frac{2}{3}\right)$ _____ $\mathbf{-\frac{1}{4}}$

13. $2\frac{2}{3} \div \frac{1}{2}$ _____ $\mathbf{5\frac{1}{3}}$

14. $1\frac{1}{4} \div \frac{1}{6}$ _____ $\mathbf{7\frac{1}{2}}$

15. $2\frac{1}{2} \div \frac{7}{8}$ _____ $\mathbf{2\frac{6}{7}}$

16. $2\frac{1}{2} \div 3\frac{1}{2}$ _____ $\mathbf{\frac{5}{7}}$

17. $1\frac{1}{6} \div 1\frac{1}{3}$ _____ $\mathbf{\frac{7}{8}}$

18. $\left(-1\frac{1}{5}\right) \div 2\frac{1}{5}$ _____ $\mathbf{-\frac{6}{11}}$

19. A jug holds 12 pints of juice. How many $1\frac{1}{2}$-pint bottles can be filled with that much juice?

   8 bottles

Holt Mathematics

Holt Mathematics

LESSON 3-11
Practice B
Dividing Fractions and Mixed Numbers

Divide. Write each answer in simplest form.

1. $4 \div \frac{1}{2}$

8

2. $\frac{1}{5} \div \frac{1}{4}$

$\frac{4}{5}$

3. $\frac{1}{3} \div \frac{3}{5}$

$\frac{5}{9}$

4. $\frac{8}{9} \div \frac{2}{3}$

$1\frac{1}{3}$

5. $-\frac{3}{8} \div \frac{3}{4}$

$-\frac{1}{2}$

6. $\frac{7}{10} \div \frac{3}{5}$

$1\frac{1}{6}$

7. $\frac{5}{12} \div \frac{2}{5}$

$1\frac{1}{24}$

8. $\frac{3}{4} \div \frac{4}{9}$

$1\frac{11}{16}$

9. $\frac{7}{12} \div \frac{3}{4}$

$\frac{7}{9}$

10. $-4\frac{1}{6} \div \frac{1}{3}$

$-12\frac{1}{2}$

11. $3\frac{1}{4} \div \frac{2}{5}$

$8\frac{1}{8}$

12. $6\frac{1}{9} \div \frac{1}{6}$

$36\frac{2}{3}$

13. $2\frac{1}{4} \div 1\frac{3}{4}$

$1\frac{2}{7}$

14. $3\frac{3}{4} \div 2\frac{5}{6}$

$1\frac{11}{34}$

15. $5\frac{1}{3} \div -1\frac{4}{5}$

$-2\frac{26}{27}$

16. $2\frac{1}{2} \div 2\frac{1}{3}$

$1\frac{1}{14}$

17. $-1\frac{3}{4} \div 1\frac{1}{4}$

$-1\frac{2}{5}$

18. $7\frac{2}{3} \div 1\frac{1}{5}$

$6\frac{7}{18}$

19. Burger Barn has $46\frac{2}{3}$ pounds of ground beef. How many $\frac{1}{3}$-pound burgers can be made using all the ground beef?

140 burgers

20. Roberto needs some roofing tiles to be cut from a large tile. How many tiles that are each $14\frac{3}{8}$ inches in length can he cut from a larger piece of tile that is $100\frac{5}{8}$ inches long?

7 tiles

LESSON 3-11
Practice C
Dividing Fractions and Mixed Numbers

Divide. Write each answer in simplest form.

1. $17 \div \frac{4}{9}$

$38\frac{1}{4}$

2. $23 \div \frac{5}{7}$

$32\frac{1}{5}$

3. $39 \div \frac{7}{11}$

$61\frac{2}{7}$

4. $-\frac{11}{15} \div \frac{7}{9}$

$-\frac{33}{35}$

5. $\frac{9}{14} \div \frac{11}{21}$

$1\frac{5}{22}$

6. $\frac{15}{19} \div \frac{3}{8}$

$2\frac{2}{19}$

7. $\frac{3}{25} \div \frac{11}{15}$

$\frac{9}{55}$

8. $\frac{5}{39} \div -\frac{5}{13}$

$-\frac{1}{3}$

9. $\frac{7}{17} \div \frac{14}{17}$

$\frac{1}{2}$

10. $5\frac{3}{7} \div \frac{4}{9}$

$12\frac{3}{14}$

11. $7\frac{8}{9} \div \frac{5}{11}$

$17\frac{16}{45}$

12. $-16\frac{2}{11} \div \frac{5}{9}$

$-19\frac{23}{55}$

13. $8\frac{5}{7} \div 3\frac{6}{7}$

$2\frac{7}{27}$

14. $9\frac{3}{8} \div 2\frac{7}{8}$

$3\frac{6}{23}$

15. $5\frac{3}{5} \div 1\frac{3}{8}$

$4\frac{4}{55}$

16. $-\frac{2}{9} \div \frac{7}{8} \div \frac{1}{3}$

$-\frac{16}{21}$

17. $\frac{3}{5} \div 4\frac{1}{4} \div \frac{2}{5}$

$\frac{6}{17}$

18. $3\frac{2}{3} \div \frac{2}{7} \div \frac{7}{9}$

$16\frac{1}{2}$

19. Jorge can mow 1 lawn in $\frac{3}{4}$ hour. How many lawns can he mow in $3\frac{3}{4}$ hours?

5 lawns

20. How many $2\frac{1}{2}$-foot shelves can you cut from a $17\frac{1}{2}$-foot board?

7 shelves

LESSON 3-11
Reteach
Dividing Fractions and Mixed Numbers

Dividing fractions and mixed numbers is very much like multiplying fractions and mixed numbers. Just follow these steps:

> **Step 1:** Write any mixed numbers as improper fractions.
> **Step 2:** Invert the divisor.
> **Step 3:** Multiply and write the quotient in simplest form.

Divide: $1\frac{1}{8} \div \frac{1}{3}$

Step 1: $1\frac{1}{8} \div \frac{1}{3} = \frac{9}{8} \div \frac{1}{3}$

Step 2: $\frac{9}{8} \div \frac{1}{3} = \frac{9}{8} \cdot \frac{3}{1}$

Step 3: $\frac{9}{8} \cdot \frac{3}{1} = \frac{27}{8} = 3\frac{3}{8}$

Divide: $1\frac{1}{4} \div 3\frac{1}{3}$

Step 1: $1\frac{1}{4} \div 3\frac{1}{3} = \frac{5}{4} \div \frac{10}{3}$

Step 2: $\frac{5}{4} \div \frac{10}{3} = \frac{5}{4} \cdot \frac{3}{10}$

Step 3: $\frac{5}{4} \cdot \frac{3}{10} = \frac{15}{40} = \frac{3}{8}$

Divide. Write each answer in simplest form.

1. $\frac{4}{5} \div \frac{1}{2} = \frac{4}{5} \cdot \frac{2}{1} = \frac{8}{5} = 1\frac{3}{5}$

2. $\frac{5}{8} \div \frac{5}{6} = \frac{5}{8} \cdot \frac{6}{5} = \frac{30}{40} = \frac{3}{4}$

3. $2\frac{1}{2} \div 1\frac{3}{4} = \frac{5}{2} \div \frac{7}{4} = \frac{5}{2} \cdot \frac{4}{7}$

$= \frac{20}{14} = \frac{10}{7} = 1\frac{3}{7}$

4. $2\frac{2}{3} \div 1\frac{1}{5} = \frac{8}{3} \div \frac{6}{5} = \frac{8}{3} \cdot \frac{5}{6}$

$= \frac{40}{18} = \frac{20}{9} = 2\frac{2}{9}$

5. $\frac{3}{5} \div \frac{3}{10}$

2

6. $\frac{7}{8} \div \frac{1}{3}$

$2\frac{5}{8}$

7. $\frac{5}{12} \div \frac{1}{2}$

$\frac{5}{6}$

8. $4\frac{1}{3} \div 1\frac{1}{9}$

$3\frac{9}{10}$

9. $2\frac{1}{3} \div 1\frac{3}{4}$

$1\frac{1}{3}$

10. $5\frac{5}{8} \div 2\frac{1}{2}$

$2\frac{1}{4}$

LESSON 3-11
Challenge
Animal Senior Citizens

The value of each division expression will show the average life expectancy of an animal. Complete the chart. Then use the information to fill in the blanks below.

	Animal	Division Expression	Life Expectancy (yr)
1.	Black bear	$2\frac{1}{2} \div \frac{5}{36}$	18
2.	Chimpanzee	$40\frac{2}{5} \div 2\frac{1}{50}$	20
3.	Dog	$15\frac{1}{2} \div 1\frac{7}{24}$	12
4.	Hippopotamus	$4\frac{3}{8} \div \frac{7}{64}$	40
5.	Mouse	$\frac{5}{6} \div \frac{5}{18}$	3
6.	Guinea pig	$\frac{2}{3} \div \left(-\frac{3}{5}\right) \div \left(-\frac{5}{18}\right)$	4
7.	Tiger	$-2\frac{1}{4} \div 3\frac{1}{2} \div \left(-\frac{9}{224}\right)$	16
8.	Kangaroo	$\left(\frac{3}{4} + \frac{7}{8}\right) \div \frac{13}{56}$	7

9. If you divide the average life span of a guinea pig by $\frac{1}{10}$, you get the average life span of a hippopotamus.

10. If you divide the average life span of a kangaroo by $\frac{7}{20}$, you find the average life span of a horse, which is __20 years__.

11. A __hippopotamus__ can live 37 years longer than a __mouse__. This is $5\frac{1}{2} \div 2\frac{3}{4}$ years longer than the life expectancy of an African elephant, which is __38 years__.

12. (Average life expectancy of a dog) $\div \frac{3}{8} \div 6\frac{2}{5}$ = (average life expectancy of a rabbit) = __5 years__.

Problem Solving
Dividing Fractions and Mixed Numbers

Write the correct answer.

1. The Wheeling Bridge in West Virginia is about $307\frac{4}{5}$ meters long. If you walk with a stride of about $\frac{3}{10}$ meter, how many steps would it take you to cross this suspension bridge?

about 1,026 steps

2. The Flathead Rail Tunnel in Montana is about $7\frac{3}{4}$ miles long. If a train travels through the tunnel at a speed of about $1\frac{1}{2}$ miles per minute, how long will it take to pass from one end of the tunnel to the other?

$5\frac{1}{6}$ **minutes**

3. A hiking trail is $6\frac{2}{3}$ miles long. It has 4 exercise stations, spaced evenly along the trail. What is the distance between each exercise station?

$1\frac{2}{3}$ **miles**

4. Jamal buys a strip of 25 postage stamps. The strip of stamps is $21\frac{7}{8}$ inches long. How long is each stamp?

$\frac{7}{8}$ **inch**

Choose the letter for the best answer.

5. Matt wants to decorate his skateboard with decals. His skateboard is $28\frac{3}{4}$ inches long. The decals are $5\frac{1}{2}$ inches long. If Matt arranges them in a line from the front to the back, how many decals will fit?
   Ⓐ 5 decals
   B 6 decals
   C 7 decals
   D 8 decals

6. A square floor tile measures $\frac{3}{4}$ square feet. How many tiles are required to cover a 200 square foot floor?
   F 150 tiles
   Ⓖ 267 tiles
   H 275 tiles
   J 300 tiles

7. Bev buys a sleeve of ball bearings for her skateboard. Each of the bearings is $1\frac{1}{5}$ inches wide. The sleeve is $9\frac{3}{5}$ inches long. How many bearings are in the sleeve?
   A 5 bearings
   Ⓑ 8 bearings
   C 9 bearings
   D 12 bearings

8. The average hamster weighs about $\frac{1}{4}$ pound. The total weight of all the hamsters in a cage in a pet store is $1\frac{1}{2}$ pounds. How many hamsters are in the cage?
   F 3 hamsters
   G 4 hamsters
   H 5 hamsters
   Ⓙ 6 hamsters

91
Holt Mathematics

Reading Strategies
Use a Visual Model

The Smith family has a two-and-a-half-foot-long sandwich to share. One-half foot of the sandwich will serve one person. How many one-half foot servings are in this sandwich?

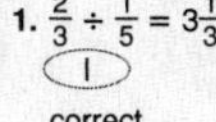

Use the model to answer each question.

1. How long is the sandwich?
   $2\frac{1}{2}$ ft

2. How long is each serving?
   $\frac{1}{2}$ ft

3. If you divided the sandwich into $\frac{1}{2}$ ft servings, how many would you have?
   5

4. What is $2\frac{1}{2} \div \frac{1}{2}$?
   5

Suppose you have two sandwiches.

5. How many feet are in both sandwiches?
   5

6. What is $2\frac{1}{2} \times 2$?
   5

7. Compare the answers to $2\frac{1}{2} \div \frac{1}{2}$ and $2\frac{1}{2} \times 2$. What do you notice?
   The answers are the same.

92
Holt Mathematics

Puzzles, Twisters & Teasers
Stop Making Sense!

Decide whether or not each equation is correct. Circle the letter above your answer. Use the letters you circled to solve the riddle.

1. $\frac{2}{3} \div \frac{1}{5} = 3\frac{1}{3}$
   (I) correct   B incorrect

2. $2 \div -\frac{7}{8} = 4\frac{1}{4}$
   D correct   (T) incorrect

3. $\frac{3}{4} \div \frac{6}{7} = \frac{1}{3}$
   F correct   (M) incorrect

4. $3\frac{3}{5} \div 9\frac{1}{7} = -3\frac{3}{5}$
   G correct   (A) incorrect

5. $-4\frac{1}{3} \div 2\frac{1}{2} = -1\frac{11}{15}$
   (K) correct   H incorrect

6. $\frac{3}{5} \div 6 = \frac{1}{10}$
   (E) correct   J incorrect

7. $9 \div 1\frac{1}{2} = 6$
   (S) correct   L incorrect

8. $10 \div \frac{5}{9} = \frac{3}{10}$
   O correct   (C) incorrect

9. $-4\frac{4}{7} \div \frac{6}{7} = -5\frac{3}{5}$
   (E) correct   P incorrect

10. $\frac{2}{3} \div \frac{8}{9} = 1$
    Q correct   (N) incorrect

11. $\frac{3}{4} \div \frac{6}{7} = \frac{6}{8}$
    R correct   (T) incorrect

12. $\frac{4}{9} \div 12 = 1\frac{3}{4}$
    V correct   (S) incorrect

What did the little boy say when he learned how to count money?

I T
M A K E S
C E N T S !

93
Holt Mathematics

Practice A
Solving Equations Containing Fractions

Solve. Choose the letter for the best answer.

1. $t - \frac{3}{4} = \frac{1}{4}$
   A $t = \frac{1}{4}$ **C** $t = \frac{3}{4}$
   B $t = \frac{1}{2}$ **Ⓓ** $t = 1$

2. $g - \frac{3}{8} = \frac{1}{8}$
   F $g = \frac{1}{4}$ **H** $g = \frac{3}{4}$
   Ⓖ $g = \frac{1}{2}$ **J** $g = 1$

3. $k + \frac{7}{12} = \frac{11}{12}$
   A $k = \frac{1}{4}$ **C** $k = \frac{1}{2}$
   Ⓑ $k = \frac{1}{3}$ **D** $k = 1$

4. $n + \frac{2}{5} = \frac{4}{5}$
   Ⓕ $n = \frac{2}{5}$ **H** $n = \frac{4}{5}$
   G $n = \frac{3}{5}$ **J** $n = 1$

5. $f + \frac{1}{6} = \frac{5}{6}$
   A $f = \frac{1}{6}$ **C** $f = \frac{1}{2}$
   B $f = \frac{1}{3}$ **Ⓓ** $f = \frac{2}{3}$

6. $\frac{1}{4}s = 4$
   F $s = \frac{1}{2}$ **H** $s = 4$
   G $s = 1$ **Ⓙ** $s = 16$

7. $\frac{2}{5}a = 10$
   A $a = 5$ **Ⓒ** $a = 25$
   B $a = 10$ **D** $a = 50$

8. $\frac{1}{6}t = \frac{1}{2}$
   F $t = 1$ **H** $t = 6$
   Ⓖ $t = 3$ **J** $t = 12$

Solve. Write each answer in simplest form.

9. $p - \frac{1}{4} = \frac{1}{6}$
   $p = \frac{5}{12}$

10. $d - \frac{2}{5} = \frac{3}{10}$
    $d = \frac{7}{10}$

11. $y + \frac{5}{8} = \frac{3}{4}$
    $y = \frac{1}{8}$

12. $\frac{3}{4}m = \frac{5}{6}$
    $m = 1\frac{1}{9}$

13. $\frac{1}{2}x = \frac{5}{8}$
    $x = 1\frac{1}{4}$

14. $\frac{5}{6}r = \frac{3}{10}$
    $r = \frac{9}{25}$

15. Eunice walked $\frac{1}{4}$ mile from home to the store on her way to a friend's house. If the store is $\frac{1}{3}$ of the way to her friend's house, how far is her friend's house from home?
    $\frac{3}{4}$ mile

94
Holt Mathematics

Practice B
Solving Equations Containing Fractions

Solve. Write each answer in simplest form.

1. $t - \frac{3}{7} = \frac{4}{7}$

$t = 1$

2. $g - \frac{5}{16} = \frac{3}{16}$

$g = \frac{1}{2}$

3. $k - \frac{3}{10} = \frac{2}{5}$

$k = \frac{7}{10}$

4. $n + \frac{1}{7} = \frac{2}{3}$

$n = \frac{11}{21}$

5. $j + \frac{5}{6} = \frac{17}{18}$

$j = \frac{1}{9}$

6. $f + \frac{5}{12} = \frac{3}{4}$

$f = \frac{1}{3}$

7. $\frac{1}{4}s = \frac{3}{4}$

$s = 3$

8. $\frac{1}{5}a = \frac{1}{2}$

$a = 2\frac{1}{2}$

9. $\frac{4}{5}h = \frac{8}{9}$

$h = 1\frac{1}{9}$

10. $p - \frac{2}{3} = \frac{5}{8}$

$p = 1\frac{7}{24}$

11. $d - \frac{3}{5} = \frac{7}{10}$

$d = 1\frac{3}{10}$

12. $y - \frac{2}{7} = 3\frac{1}{4}$

$y = 3\frac{15}{28}$

13. $c + \frac{5}{12} = 2\frac{1}{6}$

$c = 1\frac{3}{4}$

14. $w + \frac{4}{15} = 3\frac{1}{3}$

$w = 3\frac{1}{15}$

15. $z + \frac{6}{7} = 2\frac{3}{5}$

$z = 1\frac{26}{35}$

16. $\frac{5}{6}m = \frac{8}{9}$

$m = 1\frac{1}{15}$

17. $\frac{1}{2}x = 3\frac{7}{15}$

$x = 6\frac{14}{15}$

18. $\frac{1}{5}r = 2\frac{2}{3}$

$r = 13\frac{1}{3}$

19. Sarabeth ran $1\frac{2}{5}$ miles on a path around the park. This was $\frac{5}{8}$ of the distance around the park. What is the distance around the park?

$2\frac{6}{25}$ miles

20. An interior decorator bought $12\frac{1}{2}$ yards of material to make drapes. He used $8\frac{2}{3}$ yards on 1 pair of drapes. How much material does he have left?

$3\frac{5}{6}$ yards

Practice C
Solving Equations Containing Fractions

Solve. Write each answer in simplest form.

1. $t - \frac{12}{19} = \frac{5}{19}$

$t = \frac{17}{19}$

2. $g - \frac{5}{22} = \frac{3}{11}$

$g = \frac{1}{2}$

3. $k - \frac{3}{4} = \frac{2}{3}$

$k = 1\frac{5}{12}$

4. $n + \frac{8}{15} = \frac{3}{10}$

$n = -\frac{7}{30}$

5. $j + \frac{11}{12} = \frac{13}{18}$

$j = -\frac{7}{36}$

6. $f + \frac{8}{11} = \frac{4}{5}$

$f = \frac{4}{55}$

7. $\frac{1}{7}s = 1\frac{1}{4}$

$s = 8\frac{3}{4}$

8. $\frac{3}{5}a = 3\frac{2}{7}$

$a = 5\frac{10}{21}$

9. $-\frac{4}{9}h = -6\frac{2}{7}$

$h = 14\frac{1}{7}$

10. $p - \frac{2}{7} = 2\frac{3}{8}$

$p = 2\frac{37}{56}$

11. $d + \frac{5}{12} = -2\frac{11}{15}$

$d = -2\frac{11}{20}$

12. $y - \frac{2}{5} = 4\frac{4}{9}$

$y = 4\frac{38}{45}$

13. $-\frac{7}{10} + c = 5\frac{3}{20}$

$c = -5\frac{17}{20}$

14. $w + 1\frac{7}{11} = 4\frac{2}{5}$

$w = 2\frac{42}{55}$

15. $z + 3\frac{1}{6} = 6\frac{2}{9}$

$z = 3\frac{1}{18}$

16. $\frac{5}{8}m = 2\frac{4}{7}$

$m = 4\frac{4}{35}$

17. $-\frac{1}{11}x = 2\frac{1}{10}$

$x = -23\frac{1}{10}$

18. $\frac{3}{7}r = 3\frac{1}{3}$

$r = 7\frac{7}{9}$

19. A grizzly bear can run $1\frac{1}{5}$ times as fast as an elephant. A grizzly bear can run 30 mi/h. How fast can an elephant run?

25 mi/h

20. A fruit salad recipe with bananas, strawberries, and grapes requires $3\frac{1}{2}$ pounds of fruit. You need $1\frac{3}{4}$ pounds of bananas and $\frac{7}{8}$ of a pound of strawberries. How many pounds of grapes are needed?

$\frac{7}{8}$ pound

Reteach
Solving Equations Containing Fractions

You can use addition to solve a subtraction equation involving fractions.

$$x - \frac{4}{9} = \frac{1}{3}$$
$$x - \frac{4}{9} + \frac{4}{9} = \frac{1}{3} + \frac{4}{9}$$
$$x = \frac{3}{9} + \frac{4}{9}$$
$$x = \frac{7}{9}$$

Remember, addition undoes subtraction.

You can use subtraction to solve an addition equation involving fractions.

$$n + \frac{2}{5} = \frac{9}{10}$$
$$n + \frac{2}{5} - \frac{2}{5} = \frac{9}{10} - \frac{2}{5}$$
$$n = \frac{9}{10} - \frac{4}{10}$$
$$n = \frac{5}{10} = \frac{1}{2}$$

Remember, subtraction undoes addition.

Solve. Write each answer in simplest form.

1. $d - \frac{1}{6} = \frac{3}{4}$

$$d - \frac{1}{6} + \frac{1}{6} = \frac{3}{4} + \frac{1}{6}$$
$$d = \frac{9}{12} + \frac{2}{12}$$
$$d = \frac{11}{12}$$

2. $y + \frac{4}{5} = \frac{14}{15}$

$$y + \frac{4}{5} - \frac{4}{5} = \frac{14}{15} - \frac{4}{5}$$
$$y = \frac{14}{15} - \frac{12}{15}$$
$$y = \frac{2}{15}$$

3. $t - \frac{1}{8} = \frac{3}{4}$

$t = \frac{7}{8}$

4. $k + \frac{1}{2} = 1\frac{5}{8}$

$k = 1\frac{1}{8}$

5. $a - \frac{3}{5} = \frac{7}{10}$

$a = 1\frac{3}{10}$

Reteach
Solving Equations Containing Fractions (continued)

You can use division to solve a multiplication equation involving fractions. Multiply both sides of the equation by the reciprocal of the coefficient of the variable.

$$3y = \frac{9}{10}$$
$$3y \cdot \frac{1}{3} = \frac{9}{10} \cdot \frac{1}{3}$$
$$y = \frac{9}{10} \cdot \frac{1}{3}$$
$$y = \frac{9}{30} = \frac{3}{10}$$

The reciprocal of 3 is $\frac{1}{3}$.

$$\frac{5}{6}a = \frac{1}{2}$$
$$\frac{5}{6}a \cdot \frac{6}{5} = \frac{1}{2} \cdot \frac{6}{5}$$
$$a = \frac{1}{2} \cdot \frac{6}{5}$$
$$a = \frac{6}{10} = \frac{3}{5}$$

The reciprocal of $\frac{5}{6}$ is $\frac{6}{5}$.

Solve. Write each answer in simplest form.

6. $8x = 3\frac{1}{5}$

$$8x = \frac{16}{5}$$
$$8x \cdot \frac{1}{8} = \frac{16}{5} \cdot \frac{1}{8}$$
$$x = \frac{16}{5} \cdot \frac{1}{8}$$
$$x = \frac{16}{40} = \frac{2}{5}$$

7. $\frac{2}{3}k = \frac{5}{6}$

$$\frac{2}{3}k \cdot \frac{3}{2} = \frac{5}{6} \cdot \frac{3}{2}$$
$$k = \frac{5}{6} \cdot \frac{3}{2}$$
$$k = \frac{15}{12} = \frac{5}{4} = 1\frac{1}{4}$$

8. $\frac{3}{4}d = 5$

$d = 6\frac{2}{3}$

9. $6y = \frac{2}{3}$

$y = \frac{1}{9}$

10. $\frac{1}{5}s = \frac{5}{8}$

$s = 3\frac{1}{8}$

Challenge
Shopping List Equation

On the right is the list of ingredients for a recipe for bean and cheese tacos. The recipe serves 4 people. You are making bean and cheese tacos for 18 people.

Use the ingredients to answer the questions and solve the equations.

Bean and Cheese Tacos
$\frac{1}{2}$ pound kidney beans
1 clove garlic, chopped
4 flour tortillas
1 cup ricotta cheese
$\frac{1}{4}$ cup parmesan cheese
$\frac{1}{4}$ cup green onions, chopped
Serves 4

1. By what factor do you need to multiply each ingredient to serve 18 people? Solve the equation to find out. $4x = 18$

$x = 4\frac{1}{2}$

2. How many pounds of kidney beans do you need to serve 18 people?

$2\frac{1}{4}$ pounds

3. Kidney beans come in $\frac{3}{4}$-pound cans. How many cans do you need to serve 18 people? Write and solve an equation.

$\frac{3}{4}x = 2\frac{1}{4}$; $x = 3$ cans

4. Ricotta cheese is on sale in $1\frac{1}{4}$-cup tubs. How many tubs do you need to serve 18 people? Write and solve an equation.

$1\frac{1}{4}x = 4\frac{1}{2}$; $x = 3\frac{3}{5}$; You need 4 tubs.

5. You need $1\frac{1}{8}$ cups of Parmesan cheese. You have $\frac{1}{4}$ cup. Write and solve an equation to find how much more you need.

$\frac{1}{4} + x = 1\frac{1}{8}$; $x = \frac{7}{8}$ cup

6. There are 4 green onions in $\frac{1}{4}$ cup. Write and solve an equation to find how many green onions you need.

$4\frac{1}{2} = \frac{1}{4}x$; $x = 18$ green onions

7. You have 8 cloves of garlic and you need $4\frac{1}{2}$. How many cloves of garlic will you have left? Write and solve an equation.

$8 - x = 4\frac{1}{2}$; $x = 3\frac{1}{2}$ cloves left

Holt Mathematics

Problem Solving
Solving Equations Containing Fractions

Write the correct answer.

1. At the 2002 Winter Olympics, Austria won 2 gold medals. This was $\frac{1}{8}$ of the total medals Austria won. How many medals did Austria win?

__16 medals__

2. At the 2002 Winter Olympics, Germany won 35 medals, of which 16 were silver. They won $1\frac{1}{3}$ times as many silver medals as gold medals. How many gold medals did Germany win?

__12 gold medals__

3. Jesse plays ice hockey. He is on the ice $\frac{2}{5}$ of each game. A game lasts 45 minutes, divided into 3 periods. How many minutes is Jesse on the ice during each game?

__18 minutes__

4. Amelia's soccer team won $\frac{3}{4}$ of its games. The team won 18 games. How many games did the team play?

__24 games__

Choose the letter for the best answer.

5. At the 2002 Winter Olympics, China won 8 medals. The United States won $4\frac{1}{4}$ times as many medals as China did. How many medals did the United States win?

A 24 medals   C 34 medals
B 32 medals   D 40 medals

6. Juanita's water bottle contains $1\frac{7}{8}$ liters. She uses her bottle to fill Felix's. When she is done, her bottle contains $\frac{1}{4}$ liter. How many liters can Felix's bottle hold?

F $\frac{7}{8}$ liters   H $1\frac{5}{8}$ liters
G $1\frac{1}{8}$ liters   J $2\frac{1}{8}$ liters

7. Yoriko ran $6\frac{1}{2}$ laps of the track. Each lap is $\frac{1}{4}$ of a mile. How many miles did she run?

A $6\frac{5}{8}$ miles   C 2 miles
B $4\frac{1}{2}$ miles   D $1\frac{5}{8}$ miles

8. Rocky runs $3\frac{1}{2}$ miles each week. Leroy runs $5\frac{1}{3}$ miles each week. How much farther does Leroy run?

F $1\frac{1}{2}$ miles   H $2\frac{1}{6}$ miles
G $1\frac{5}{6}$ miles   J $2\frac{1}{4}$ miles

Holt Mathematics

Reading Strategies
Compare and Contrast

Compare the steps for solving equations with fractions and solving equations with whole numbers.

Steps for Solving Equations	Whole Numbers	Fractions
Step 1: Get x by itself on one side of the equation.	$x - 8 = 7$	$x - \frac{3}{12} = \frac{4}{12}$
Step 2: Perform the opposite operation. In a subtraction problem, you add to get x by itself.	$x - 8 + 8 = 7 + 8$	$x - \frac{3}{12} + \frac{3}{12} = \frac{4}{12} + \frac{3}{12}$
Step 3: Solve.	$x = 15$	$x = \frac{7}{12}$

Use the chart to answer each question.

1. What is the first step to solve an equation with whole numbers?

__Get x by itself on one side of the equation.__

2. Compare the first step in solving an equation with whole numbers to fractions. Is it the same or different?

__the same__

3. What is the second step in solving an equation with whole numbers?

__Perform the opposite operation.__

4. Compare the second step in solving an equation with whole numbers to an equation with fractions. Is it the same or different?

__the same__

5. What is the opposite operation in both of the examples in the chart?

__addition__

6. What is the third step in solving an equation with whole numbers?

__Solve.__

7. Compare the third step in solving an equation with whole numbers to solving an equation with fractions. Is it the same or different?

__the same__

Holt Mathematics

Puzzles, Twisters & Teasers
What's Up, Doc?

Solve the equations. Then use the letters of the variables to answer the riddle.

1. $\frac{7}{18} + t = -\frac{4}{7}$ $t = -\frac{121}{126}$

2. $e - \frac{1}{5} = \frac{3}{5}$ $e = \frac{4}{5}$

3. $g + \frac{11}{20} = \frac{3}{4}$ $g = \frac{1}{5}$

4. $\frac{2}{3}k = \frac{4}{5}$ $k = 1\frac{1}{5}$

5. $3a = \frac{6}{7}$ $a = \frac{2}{7}$

6. $\frac{5}{12} + o = \frac{2}{3}$ $o = \frac{1}{4}$

7. $m - \frac{1}{2} = \frac{1}{4}$ $m = \frac{3}{4}$

8. $\frac{1}{5}r = 8$ $r = 40$

9. $s - \frac{2}{3} = \frac{5}{6}$ $s = 1\frac{1}{2}$

10. $\frac{2}{3}x = \frac{3}{5}$ $x = \frac{9}{10}$

Why did the rabbit buy a pure gold ring?

$\underset{\frac{121}{126}}{T}\ \underset{\frac{1}{4}}{O}\quad \underset{\frac{1}{5}}{G}\ \underset{\frac{4}{5}}{E}\ \underset{\frac{121}{126}}{T}$

$\underset{\frac{3}{4}}{M}\ \underset{\frac{1}{4}}{O}\ \underset{40}{R}\ \underset{\frac{4}{5}}{E}$

$\underset{1\frac{1}{5}}{K}\ \underset{\frac{2}{7}}{A}\ \underset{40}{R}\ \underset{\frac{2}{7}}{A}\ \underset{-\frac{121}{126}}{T}\ \underset{1\frac{1}{2}}{S}$

Holt Mathematics

Holt Mathematics